Analog Circuits and Signal Processing

Series Editors:

Mohammed Ismail, Dublin, USA

Mohamad Sawan, Montreal, Canada

The Analog Circuits and Signal Processing book series, formerly known as the Kluwer International Series in Engineering and Computer Science, is a high level academic and professional series publishing research on the design and applications of analog integrated circuits and signal processing circuits and systems. Typically per year we publish between 5–15 research monographs, professional books, handbooks, edited volumes and textbooks with worldwide distribution to engineers, researchers, educators, and libraries.

The book series promotes and expedites the dissemination of new research results and tutorial views in the analog field. There is an exciting and large volume of research activity in the field worldwide. Researchers are striving to bridge the gap between classical analog work and recent advances in very large scale integration (VLSI) technologies with improved analog capabilities. Analog VLSI has been recognized as a major technology for future information processing. Analog work is showing signs of dramatic changes with emphasis on interdisciplinary research efforts combining device/circuit/technology issues. Consequently, new design concepts, strategies and design tools are being unveiled.

Topics of interest include:

Analog Interface Circuits and Systems;

Data converters;

Active-RC, switched-capacitor and continuous-time integrated filters;

Mixed analog/digital VLSI;

Simulation and modeling, mixed-mode simulation;

Analog nonlinear and computational circuits and signal processing;

Analog Artificial Neural Networks/Artificial Intelligence;

Current-mode Signal Processing; Computer-Aided Design (CAD) tools;

Analog Design in emerging technologies (Scalable CMOS, BiCMOS, GaAs, heterojunction and floating gate technologies, etc.);

Analog Design for Test;

Integrated sensors and actuators; Analog Design Automation/Knowledge-based Systems; Analog VLSI cell libraries; Analog product development; RF Front ends, Wireless communications and Microwave Circuits;

Analog behavioral modeling, Analog HDL.

More information about this series at http://www.springer.com/series/7381

Hao Gao • Marion Matters-Kammerer
Dusan Milosevic • Peter G.M. Baltus

Batteryless mm-Wave Wireless Sensors

Springer

Hao Gao
Eindhoven University of Technology
Eindhoven, The Netherlands

Marion Matters-Kammerer
Eindhoven University of Technology
Eindhoven, The Netherlands

Dusan Milosevic
Eindhoven University of Technology
Eindhoven, The Netherlands

Peter G.M. Baltus
Eindhoven University of Technology
Eindhoven, The Netherlands

ISSN 1872-082X   ISSN 2197-1854   (electronic)
Analog Circuits and Signal Processing
ISBN 978-3-319-89226-9   ISBN 978-3-319-72980-0   (eBook)
https://doi.org/10.1007/978-3-319-72980-0

Printed on acid-free paper

This Springer imprint is published by Springer Nature
The registered company is Springer International Publishing AG
The registered company address is: Gewerbestrasse 11, 6330 Cham, Switzerland

Contents

List of Abbreviations

AC	Alternating current
ADC	Analog-to-digital convertor
AoC	Antenna on chip
BAN	Body area network
BER	Bit error rate
BW	Bandwidth
CG	Common gate
CMOS	Complementary metal oxide semiconductor
CS	Common source
DAC	Digital-to-analog convertor
DC	Direct current
EBM	End-of-burst monitor
EIRP	Equivalent isotropic radiation power
EVM	Error vector magnitude
FCC	Federal communications commission
GaAs	Gallium arsenide
HF	High frequency
IC	Integrated circuit
IIP3	Input-referred third-order intercept point
IJLO	Injection locked oscillator
IR-UWB	Impulse radio ultra-wideband
ISM	Industrial scientific medical
LF	Low frequency
LNA	Low noise amplifier
LO	Local oscillator
LPF	Low pass filter
MiM	Metal insulator metal

MMID	mm-Wave identification
mm-Wave	Millimeter wave
MoM	Metal oxide metal
NEF	Noise excess factor
NF	Noise figure
OOK	On off keying
PA	Power amplifier
PREMISS	Power reduced monolithic wireless sensor networks
RF	Radio frequency
RFID	Radio frequency identification
RTWU	Radio triggered wake-up
Rx	Receiver
SiGe	Silicon germanium
SNR	Signal-to-noise ratio
SPDT	Single pole double throw
SPICE	Simulation program with integrated circuit emphasis
SPST	Signal pole single throw
TE	Transverse electric
Tx	Transmitter
UHF	Ultra high frequency
ULP	Ultra low power
UWB	Ultra wide band
VCO	Voltage controlled oscillator
VGA	Variable gain amplifier
WPT	Wireless power transfer
WSN	Wireless sensor network
WuRx	Wake-up receiver

Chapter 1
Introduction

Abstract Due to their ease of deployment, wireless sensors have been used in a wide range of applications, including security, green building, automotive and biomedical. Most of the state-of-the-art wireless sensors use batteries as a power source. It can be easily calculated that for a building with 1000 wireless sensors installed, which can be very common for a smart building, assuming a battery life of 3 years, on average, batteries need to be replaced every day. Therefore, from the point of view of cost, convenience, environment, and reliability, there is a strong demand for battery-less wireless sensors.

1.1 Background

Wireless sensors need power for both sensing and communication with other sensor nodes or a base-station. One way to avoid batteries is to use energy scavenging techniques exploiting, e.g., solar, kinetic, and thermal energy to provide power for sensing and communication. In state-of-the-art battery-less sensors, energy scavenging and sensing are implemented in two separate modules. This increases both the cost and complexity of wireless sensors. Another disadvantage of this separation is that the robustness of the sensor node to, e.g., mechanical vibrations and temperature fluctuations is hampered thus limiting the application of such sensors in harsh environments or in moving objects. Also the availability of the light will limit the solar power scavenging.

Another way to avoid batteries is transferring the energy wirelessly to the sensor nodes when needed. A popular example of this method is the radio frequency identification (RFID) system. In RFID systems, the energy is transferred wirelessly from a reader to a tag to activate the tag for decoding the message from the reader and transmitting back the required information to the reader. The RFID systems started from low-frequency (LF) and high-frequency (HF) bands and moved to ultra-high-frequency (UHF) and 2.4 GHz industrial-scientific-medical (ISM) bands. Recently, we noticed an increasing number of publications for RFID in millimeter wave frequency bands, called millimeter wave identification (MMID), at 60 GHz [1] and 77 GHz [2]. There are a couple of advantages of moving into the millimeter

© Springer International Publishing AG 2018   1
H. Gao et al., *Batteryless mm-Wave Wireless Sensors*, Analog Circuits
and Signal Processing, https://doi.org/10.1007/978-3-319-72980-0_1

wave bands [1]. Firstly, at this high frequency, the antenna size is in the order of millimeters which could make the sensor nodes very small. Moreover, it is possible to implement antenna arrays at the reader to provide highly directional antenna gain and good spatial selectivity. Secondly, at millimeter wave frequencies, high data-rate communication with possible data rates in the order of Gbit/s can be achieved, thus enabling fast data transfer.

1.2 Scope of the Book

To overcome the limitations of state-of-the-art battery-less wireless sensors in size, cost, robustness, and range [3], we propose a system concept for a 60 GHz wireless sensor system with monolithic sensors. In the power reduced monolithic wireless sensor system [4], the wireless sensors consist of wireless power receiving, sensing and communication functions in a single chip. The sensors have no external components and hence avoid costly IC-interfaces that are sensitive to mechanical and thermal stress. This book focused on, but was not limited to:

- Analysis of the wireless power transfer system specifications.
- Analysis of rectifier performance and its limitations at mm-wave frequencies as a wireless power receiver. Design and implementation of new concepts in mm-wave rectifiers in order to improve the rectifier performance and realize the on-chip mm-wave wireless power receiver.
- Integration of rectifier with the on-chip antenna and ultra-low-power radio, in order to achieve a monolithic sensor node and evaluation of its performance.
- Analysis of 60 GHz ultra-low-power radio techniques; design and implementation of a 60 GHz ultra-low-power receiver in advanced CMOS technology.
- Analysis of the 60 GHz high performance wireless data transfer and power transfer for the base-station. The phased-array architecture is used for the base-station to increase the power density at the sensor nodes and increase receiver sensitivity at the base-station.
- Design, implementation, and evaluation of the phase shifters and low noise amplifiers in advanced CMOS technology.
- Design and implementation of an Rx front-end for a phased-array-based receiver.

Some boundary conditions on the scope of the book are explained below: (a) The 60 GHz band has been chosen because it is a worldwide available ISM band. But the same concept of the wireless power transfer and phased-array technology could be adapted to other bands, such as Ku/K/Ka, W, and D band. (b) This book is focused on integration into CMOS technology. This is because of the opportunity of future system integration with the digital part. CMOS technology is selected because of the low-cost solution for high volume production, and because it could be used to achieve very large scale complex digital logic co-integration. (c) In this book, both 65 and 40 nm CMOS technology are used for the RF/mm-IC components. The same design method or topology is not limited to CMOS technology, it could also be applied to other technologies, such as SiGe or GaAs. (d) This book focuses on the phased-array techniques, because it can compensate the path loss and alleviate the requirements of 60 GHz RF transceivers. A low noise amplifier and a phase shifter form the RF front-end for one element in the RF-phase shifting based phased-array architectures. Besides the RF-phase shifting, the phase shifting could be implemented in the LO path or IF path. The design considerations of the RF phase shifter could also be applied to the phase shifter located at LO/IF path. (e) This book does not aim at a specific application. The proposed design method and considerations can be of more general use, and can be applied in other mm-wave circuit design and system considerations. In this book, a temperature sensor is chosen as example, but the on-chip wireless power transfer method can be applied to other types of sensor nodes as well.

1.3 Outline of the Book

The outline of this book is briefly explained below:

- Chapter 2 provides an overview and trends for wireless systems for wireless power transfer and data transmission.
- Chapter 3 provides the system analysis of the mm-wave wirelessly powered sensor node system.
- Chapter 4 discusses the modeling of the rectifier. Two models are provided, the first one is for the input voltage swing below the threshold voltage, and the second one is for the input voltage swing larger than the threshold voltage.
- Chapter 5 focuses on the discussion of mm-wave rectifiers. Special attention is given to the efficiency of the rectifier at mm-wave frequencies, and options to increase the efficiency; these issues are worked out. Design examples of mm-wave rectifiers in 65 nm CMOS technology are presented.
- Chapter 6 presents two mm-wave sensor nodes. The mm-wave sensor nodes include on-chip antennas, a wireless power receiver, and an ultra-low-power transmitter.
- Chapter 7 presents the system architecture for a 60 GHz ultra-low-power radio. A 60 GHz ultra-low-power radio is presented, including the design method of low power LNA, passive mixer and injection-locked oscillator. Based on the ultra-low-power radio design, a single-chip receiver is implemented in 65 nm CMOS technology.
- Chapter 8 presents an RF front-end for a phased-array system, including the key building block of LNA and passive digitally controlled phase shifter.
- Finally, conclusions are presented in Chap. 9.

References

1. H. Gao, M. Matters-Kammerer, D. Milosevic, A. van Roermund, P. Baltus, A 62 GHz inductor-peaked rectifier with 7% efficiency, in *2013 IEEE Radio Frequency Integrated Circuits Symposium (RFIC)*, pp. 189–192 (2013)
2. H. Gao, M. Matters-Kammerer, P. Harpe, D. Milosevic, U. Johannsen, A. van Roermund, P. Baltus, A 71 GHz RF energy harvesting tag with 8% efficiency for wireless temperature sensors in 65 nm CMOS, in *2013 IEEE Radio Frequency Integrated Circuits Symposium (RFIC)*, pp. 403–406 (2013)
3. P. Pursula, T. Vaha-Heikkila, A. Muller, D. Neculoiu, G. Konstantinidis, A. Oja, J. Tuovinen, Millimeter-wave identification-a new short-range radio system for low-power high data-rate applications. IEEE Trans. Microwave Theory Tech. **56**(10), 2221–2228 (2008)
4. Y. Wu, J. Linnartz, H. Gao, P. Baltus, J. Bergmans, System study of a 60 GHz wireless-powered monolithic sensor system, in *2011 8th International Conference on Information, Communications and Signal Processing (ICICS)*, pp. 1–5 (2011)

Chapter 2
State of the Art

Abstract This chapter studies trends and expectations in monolithic wireless sensor system design with respect to applications, technology evolution, and system design. Problems and opportunities are analyzed. In later chapters of this book, the ultra-low-power design concept is introduced that takes advantages of the expected opportunities in order to solve the anticipated problems.

2.1 Introduction

Conventional solutions for wireless sensors are mostly battery-operated [1, 2]. Battery replacement adds significant maintenance cost and increases size of the sensor node. Wireless energy transfer is an alternative solution to power the sensor nodes. In currently deployed solutions, the energy receiver and the sensing function are implemented in two separate modules. This separation increases the size and cost of the sensor nodes. Also, it limits the robustness of the sensor node with respect to mechanical vibration and shocks, and to continuous (and sometimes extreme) temperature fluctuations. To overcome this problem, wireless power transfer to an on-chip receiver at mm-waves is proposed. This might result in a fully monolithic sensor node. Such sensor nodes have no external antenna, battery, and pins. These advantages will reduce the overall size and the system will become less sensitive to mechanical and thermal stress. As the carrier frequency increases, RF wireless power transfer becomes increasingly attractive as it can take advantage of increasingly narrower and energy-efficient pencil beams. Moreover, higher carrier frequency will reduce the size of the antenna, enabling on-chip antenna integration.

In this chapter, a short summary of the state-of-the-art and trends of wireless power transfer systems, wireless data transfer, and wireless sensor nodes is provided.

© Springer International Publishing AG 2018
H. Gao et al., *Batteryless mm-Wave Wireless Sensors*, Analog Circuits
and Signal Processing, https://doi.org/10.1007/978-3-319-72980-0_2

2.2 Wireless Power Transfer

The first experiment of energy transfer without wire in human history was performed more than a century ago. At that time, Nikola Tesla attempted to distribute ten thousand horsepowers of electricity wirelessly [3]. He did the first demonstration of wireless power transfer for lighting bulbs [3]. Recently, with the development of communication systems, wireless transmission has been widely used for transmitting information. In order to solve the issues of small size for portable devices and limited life-time of batteries, wireless power transmission has been employed widely to build low power, very short range systems such as implanted biomedical devices and radio frequency identification devices. In these applications, the power is captured by scavenging energy from electromagnetic radiation instead of from batteries.

Wireless power transfer methods can be divided into electrical coupling, inductive coupling, and RF energy transfer [4]. In electrically (capacitively) coupled systems, the reader generates a strong, high-frequency electrical field. The reader's antenna consists of a large, electrically conductive area (electrode), generally a metal foil or a metal plate. If a high-frequency voltage is applied to the electrode, a high-frequency electric field is formed between the electrode and the earth potential (ground). The electrical coupling systems are designed for ranges of less than half a meter. Inductive coupling is based on magnetic field coupling that delivers electrical energy between coils tuned to resonate at the same frequency. The electric power is carried by the magnetic field between two coils. The resonator is formed by adding a capacitance on an inductor. All of the abovementioned techniques are based on near-field wireless transmission permit by high power density and conversion efficiency. The power transmission efficiency depends on the coupling coefficient, which depends on the distance between the two coils/resonators and their layout. Therefore, they are not suitable for mobile and remote replenishment/charging. Besides, inductive coupling requires calibration and alignment of resonators in the transmitter and receiver. In contrast, RF energy transfer has no such limitation because it is a far-field energy transfer technique by using the propagated electromagnetic field as the carrier to transfer the energy. Thus RF energy transfer is suitable for powering a larger number of devices distributed in a wide area. In [5], an RFID employing inductor coupling could support energy transfer over 2 cm at 13.56 and 19.5 MHz using loop antenna structures. In [6], a magnetic resonance coupling based wireless power transfer system is provided, which could achieve energy transfer over 60 cm distance with 10% efficiency. In [7], a 19.6 m reader-to-tag communication distance is achieved by using the RF energy transfer method at 925 MHz.

2.3 mm-Wave Wireless Power Transfer

Combining highly integrated ultra-low-power mm-wave sensor nodes with wireless power transfer (WPT) to an on-chip antenna is a path towards battery-less, fully monolithically integrated, millimeter-sized sensor nodes with only a few milligram of weight [8]. Such tiny wireless sensor nodes can be applied in low-maintenance ubiquitous sensor networks as well as in miniaturized electrical capsules for local environment sensing. At 60 GHz, the antenna length can be reduced to only 0.6 mm on silicon. Furthermore, a phased-array system with beam-forming [9] applied at the base-station can make the mm-wave WPT more efficient. The rectifier is the key component to achieve the on-chip wireless power receiver. In [10] a rectifier at 45 GHz with 1.2% efficiency with 2 dBm input power has been reported. In [11], a W-band device centered at 94 GHz is reported with 10% efficiency.

2.4 Techniques for Low Power Consumption

In this section, widely used techniques and architectures for achieving low system power consumption are discussed. Those are: impulse radio ultra-wideband (IR-UWB) [12] technique, super-regenerative architecture, and wake-up radio. The IR-UWB technique could save the system power consumption from lowing SNR requirement, while the super-regenerative architecture could save the system power consumption from simplifying the system architecture. The wake-up radio introduces the idle period to save the average system power consumption.

IR-UWB is a technique for achieving a low power consumption and low cost system for short range personal communication. It can achieve low power consumption due to its low SNR requirement. In [12], an IR-UWB Tx achieves 0.65 mW power consumption with 1 Mbps data rate and 0.65 nJ/bit E_{bit}. The problem in IR-UWB is low immunity to noise and interference, the high speed and large bandwidth requirement of ADC and the difficulties from synchronization process [13]. In order to reduce the Rx power consumption, several methods are adopted in IR-UWB systems such as using simple modulation schemes (e.g., OOK), duty cycling and using a non-coherent energy detection approach. In [14], the proposed energy detector is able to achieve a sensitivity of −89 dBm and an E_{bit} of 0.2 nJ/bit with 100 kbps data rate. With a similar non-coherent detection approach, the OOK IR-UWB Rx front-end and baseband in [15] consume 1.64–2.18 mW (with 40% duty cycle) with 1 Mbps data rate, which corresponds to E_{bit} of 1.64–2.18 nJ/bit.

The super-regenerative architecture is another way to achieve low system power consumption. In this architecture, a low frequency oscillator is implemented in addition to the main oscillator. This low frequency oscillator periodically quenches the main RF oscillation to provide high non-linear gain. Although it has the drawback of poor selectivity, susceptibility to front-end overload, lack of stability and restriction to amplitude modulation schemes, it can achieve very high RF gain with very low power consumption [16]. In [17], a 402–405 MHz super-regenerative receiver is presented with 0.5 mW power consumption at a data rate of 120 Kbits/s and with a sensitivity of −95 dBm in 0.18 μm CMOS technology. In [18], an 2.36 ∼ 2.485 GHz ultra-low-power super-regenerative RF front-end is reported with 500 μW power with 1–5 Mbps data rate in 90 nm CMOS technology.

Backscattering technique is used in asymmetric systems such as RFID [19]. By adopting the self-correlation structure in the Rx [20], the requirement of Tx-Rx isolation is eliminated, and the proposed Rx is able to achieve −75 dBm sensitivity with 10 mA current consumption from a 3.3 V supply. The disadvantage of the backscattering technique is that this approach is only suitable for asymmetric wireless systems.

The wake-up receiver (WuRx) architecture is an interesting approach to decrease the system power consumption. It can reduce the stand-by power consumption of the main radio, so as to decrease the whole system power consumption. It enables a trade-off among the power consumption, data rate, sensitivity, and carrier frequency for the same BER level [21, 22]. When increasing the carrier frequency and data rate, WuRx will suffer from poor sensitivity under the same signal and power levels. However, this trade-off can be avoided by using a duty-cycled scheme with lower duty-cycle factor (DCF) as suggested in [22]. Furthermore, the minimum DCF is again limited by the detection time. In other words, if a fast-response detector can be designed at high frequencies like mm-wave, it is feasible for the WuRx to operate with ultra-low duty cycle. Consequently, a high frequency, high data rate, sensitive, low power, and "low" latency WuRx could be obtained. There are different wake-up methods, including diode detection [21], diode with simple Rx [23], uncertain-IF [24], super-regenerative [18], and multi-stage charge pump [25].

2.5   Wirelessly Powered Sensor Node

In wireless sensor networks, the requirement of supplying power to the sensor node has to be met. Using batteries for wireless sensors provides limited energy to perform demanding tasks, and how to maximize operation life-time and achieve optimal resource management remains a challenge [26]. Low battery capacity causes node malfunctions and breaks the network, and this type of WSN needs regular maintenance and battery replacement. This reduces the reliability of the WSN and increases costs. Moreover, replacing batteries introduces pollution to the environment. Charging sensor nodes remotely by an electromagnetic (EM) wave is a method to solve the battery problem for WSNs. In [27], the authors design an RF-powered transmitter that supports 915 MHz downlink and 2.45 GHz uplink bands. An average data rate of 5 kbps is achieved, while the maximum instant data rate is up to 5 Mbps. The transmitter can be operated with an input power threshold of −17.1 dBm and a maximum transmit power of −12.5 dBm. In [28], the authors present a sensor node for medical applications with dual-band operating at GSM 900 and GSM 1800. The antenna achieves gains of 1.8 ∼ 2.06 dBi with an efficiency of 77.6 ∼ 84%. In [29, 30, 31], a multi-hop RF-powered wireless sensor network is experimentally demonstrated through experiments.

2.6 Conclusion

In wirelessly powered sensor nodes, the wireless energy transfer at the mm-wave frequencies could take the advantage of minimized sensor size with on-chip antenna integration. However, the power consumption for the Rx or Tx at the mm-wave frequencies is a bottleneck. For low power mm-wave applications, the zero-IF architecture [32, 33] is preferred due to its lower power consumption compared to heterodyne receivers. In [32], a zero-IF 60 GHz receiver with BPSK modulation achieves E_{bit} of 151 pJ/bit, while its total system power consumption is 151 mW. In [33], a zero-IF 60 GHz receiver achieves E_{bit} of 151 pJ/bit with OOK modulation, and its total system power consumption is 103 mW. In [34], a 60 GHz receiver achieves 49 mW total system power consumption with envelop detection method. However, the total system power consumption is still high for achieving monolithic sensor nodes without batteries. In the work of [35], the asynchronous wake-up receiver architecture is proposed, which could provide the low-power solution for the mm-wave systems. On-chip wireless power receivers provide the opportunity to realize monolithic sensor nodes without batteries as demonstrated in the work [8]. In [10] a rectifier at 45 GHz with 1.2% efficiency with 2 dBm input power has been reported. In [11], a W-band device centered at 94 GHz is reported with 10% efficiency. Thus, the efficiency and sensitivity of the rectifier at the mm-wave frequencies should be increased to achieve monolithic sensor nodes without batteries.

References

1. N. Heidmann, N. Hellwege, D. Peters-Drolshagen, S. Paul, A. Dannies, W. Lang, A low-power wireless UHF/LF sensor network with web-based remote supervision – implementation in the intelligent container, in *2013 IEEE Sensors*, pp. 1–4 (2013)
2. A. Humbert, B. Tuerlings, R. Hoofman, Z. Tan, D. Gravesteijn, M. Pertijs, C. Bastiaansen, D. Soccol, A low-power CMOS integrated sensor for CO_2 detection in the percentage range, in *2013 Transducers Eurosensors XXVII: The 17th International Conference on Solid-State Sensors, Actuators and Microsystems (Transducers Eurosensors XXVII)*, pp. 1649–1652 (2013)
3. W.C. Brown, The history of power transmission by radio waves. IEEE Trans. Microwave Theory Tech. **32**(9), 1230–1242 (1984)
4. K. Finkenzeller, *RFID Handbook: Fundamentals and Applications in Contactless Smart Cards and Identification*, 2nd edn. (Wiley, New York, 2003)
5. O. Mourad, P. Le Thuc, R. Staraj, P. Iliev, System modeling of the RFID contactless inductive coupling using 13.56 MHz loop antennas, in *2014 8th European Conference on Antennas and Propagation (EuCAP)*, pp. 2034–2038 (2014)

6. L. Tong, H. Zeng, F. Peng, A study of the self-coupling magnetic resonance coupled wireless power transfer, in *2015 IEEE Applied Power Electronics Conference and Exposition (APEC)*, pp. 3138–3142 (2015)
7. C.-Y. Yao, W.-C. Hsia, A −21.2 dBm dual-channel UHF passive CMOS RFID tag design. IEEE Trans. Circuits Syst. Regul. Pap. **61**(4), 1269–1279 (2014)
8. H. Gao, M. Matters-Kammerer, P. Harpe, D. Milosevic, U. Johannsen, A. van Roermund, P. Baltus, A 71 GHz RF energy harvesting tag with 8% efficiency for wireless temperature sensors in 65 nm CMOS, in *2013 IEEE Radio Frequency Integrated Circuits Symposium (RFIC)*, pp. 403–406 (2013)
9. B. Wang, H. Gao, M.K. Matters-Kammerer, P.G.M. Baltus, Interpolation based wideband beamforming architecture, in *2017 IEEE International Symposium on Circuits and Systems (ISCAS)*, pp. 1–4 (2017)
10. S. Pellerano, J. Alvarado, Y. Palaskas, A mm-wave power-harvesting RFID tag in 90 nm CMOS. IEEE J. Solid State Circuits **45**(8), 1627–1637 (2010)
11. H.-K. Chiou, I.-S. Chen, High-efficiency dual-band on-chip rectenna for 35- and 94-GHz wireless power transmission in 0.13- μm CMOS technology. IEEE Trans. Microwave Theory Tech. **58**(12), 3598–3606 (2010)
12. J. Ryckaert, G. Van der Plas, V. De Heyn, C. Desset, B. Van Poucke, J. Craninckx, A 0.65-to-1.4 nJ/Burst 3-to-10 GHz UWB All-Digital TX in 90 nm CMOS for IEEE 802.15.4a. IEEE J. Solid-State Circuits **42**(12), 2860–2869 (2007)
13. H. Gao, P. Baltus, Q. Meng, 2GSPS 6-bit ADC for UWB receivers, in *2010 International Symposium on Signals, Systems and Electronics*, vol. 1, pp. 1–4 (2010)
14. A. Gerosa, S. Soldà, A. Bevilacqua, D. Vogrig, A. Neviani, An energy-detector for noncoherent impulse-radio UWB receivers. IEEE Trans. Circuits Syst. Regul. Pap. **56**(5), 1030–1040 (2009)
15. M. Crepaldi, C. Li, K. Dronson, J. Fernandes, P. Kinget, An ultra-low-power interference-robust IR-UWB transceiver chipset using self-synchronizing OOK modulation, in *2010 IEEE International Solid-State Circuits Conference Digest of Technical Papers (ISSCC)*, pp. 226–227 (2010)
16. E. Lopelli, Transceiver architectures and sub-mW fast frequency-hopping synthesizers for ultra-low power WSNs, Ph.D Dissertation, Eindhoven University of Technology (2010)
17. M. Anis, M. Ortmanns, N. Wehn, Fully integrated UWB impulse transmitter and 402-to-405 MHz super-regenerative receiver for medical implant devices, in *Proceedings of 2010 IEEE International Symposium on Circuits and Systems (ISCAS)*, pp. 1213–1215 (2010)
18. M. Vidojkovic, S. Rampu, K. Imamura, P. Harpe, G. Dolmans, H. de Groot, A 500 mW 5 Mbps ULP super-regenerative RF front-end, in *2010 Proceedings of the ESSCIRC*, pp. 462–465 (2010)
19. K. Finkenzeller, *RFID Handbook* (Wiley, New York, 2003)
20. E.-H. Kim, K. Lee, J. Ko, An isolator-less CMOS RF front-end for UHF mobile RFID reader, in *2011 IEEE International Solid-State Circuits Conference Digest of Technical Papers (ISSCC)*, pp. 456–458 (2011)
21. N. Pletcher, S. Gambini, J. Rabaey, A 65 μW, 1.9 GHz RF to digital baseband wakeup receiver for wireless sensor nodes, in *2007 IEEE Custom Integrated Circuits Conference*, pp. 539–542 (2007)
22. S. Drago, D. Leenaerts, F. Sebastiano, L. Breems, K. Makinwa, B. Nauta, A 2.4 GHz 830pJ/bit duty-cycled wake-up receiver with −82 dBm sensitivity for crystal-less wireless sensor nodes, in *2010 IEEE International Solid-State Circuits Conference Digest of Technical Papers (ISSCC)*, pp. 224–225 (2010)
23. X. Huang, S. Rampu, X. Wang, G. Dolmans, H. de Groot, A 2.4 GHz/915 MHz 51 μW wake-up receiver with offset and noise suppression, in *2010 IEEE International Solid-State Circuits Conference Digest of Technical Papers (ISSCC)*, pp. 222–223 (2010)
24. N. Pletcher, S. Gambini, J. Rabaey, A 52 μW wake-up receiver with − 72 dBm sensitivity using an uncertain-IF architecture. IEEE J. Solid-State Circuits **44**(1), 269–280 (2009)

25. L. Gu, J. Stankovic, Radio-triggered wake-up capability for sensor networks, in *Proceedings 10th IEEE Real-Time and Embedded Technology and Applications Symposium, 2004*, pp. 27–36 (2004)
26. F. Kerasiotis, A. Prayati, C. Antonopoulos, C. Koulamas, G. Papadopoulos, Battery lifetime prediction model for a WSN platform, in *2010 Fourth International Conference on Sensor Technologies and Applications (SENSORCOMM)*, pp. 525–530 (2010)
27. G. Papotto, F. Carrara, A. Finocchiaro, G. Palmisano, A 90-nm CMOS 5-Mbps crystal-less RF-powered transceiver for wireless sensor network nodes. IEEE J. Solid-State Circuits **49**(2), 335–346 (2014)
28. N. Barroca, H.M. Saraiva, P.T. Gouveia, J. Tavares, L.M. Borges, F.J. Velez, C. Loss, R. Salvado, P. Pinho, R. Goncalves, N. BorgesCarvalho, R. Chavez-Santiago, I. Balasingham, Antennas and circuits for ambient RF energy harvesting in wireless body area networks, in *2013 IEEE 24th International Symposium on Personal Indoor and Mobile Radio Communications (PIMRC)*, pp. 532–537 (2013)
29. K. Kaushik, D. Mishra, S. De, S. Basagni, W. Heinzelman, K. Chowdhury, S. Jana, Experimental demonstration of multi-hop RF energy transfer, in *2013 IEEE 24th International Symposium on Personal Indoor and Mobile Radio Communications (PIMRC)*, pp. 538–542 (2013)
30. J. Olds, W. Seah, Design of an active radio frequency powered multi-hop wireless sensor network, in *2012 7th IEEE Conference on Industrial Electronics and Applications (ICIEA)*, pp. 1721–1726 (2012)
31. W. Seah, J. Olds, Data delivery scheme for wireless sensor network powered by RF energy harvesting, in *2013 IEEE Wireless Communications and Networking Conference (WCNC)*, pp. 1498–1503 (2013)
32. A. Tomkins, R. Aroca, T. Yamamoto, S. Nicolson, Y. Doi, S. Voinigescu, A zero-IF 60 GHz 65 nm CMOS transceiver with direct BPSK modulation demonstrating up to 6 Gb/s data rates over a 2 m wireless link. IEEE J. Solid-State Circuits **44**(8), 2085–2099 (2009)
33. J. Lee, Y. Chen, Y. Huang, A low-power low-cost fully-integrated 60-GHz transceiver system with OOK modulation and on-board antenna assembly. IEEE J. Solid-State Circuits **45**(2), 264–275 (2010)
34. A. Oncu, M. Fujishima, 49 mW 5 Gbit/s CMOS receiver for 60 GHz impulse radio. Electron. Lett. **45**(17), 889–890 (2009)
35. X. Li, P. Baltus, P. van Zeijl, D. Milosevic, A. van Roermund, A 70 GHz 10.2 mW self-demodulator for OOK modulation in 65-nm CMOS technology, in *2010 IEEE Custom Integrated Circuits Conference (CICC)*, pp. 1–4 (2010)

Chapter 3
System Analysis of mm-Wave Wireless Sensor Networks

Abstract In this chapter, a battery-less wireless sensor system is proposed, which is called the Power REduced MonolithIc Sensor System (PREMISS), based on monolithic sensors with functions of wireless power receiving, wireless sensing, and wireless data transfer functions. In the PREMISS system, a base-station transmits RF energy and information to sensor nodes via pencil beams and receives the information back from those sensor nodes. In this chapter, the system analysis of the PREMISS system is presented.

3.1 Introduction

In the PREMISS system, the wireless sensors have functions of wireless power receiving, sensing and communication fully integrated into a single chip. The system also has a base-station as the central controller which is battery-operated or connected to the main, thus is much less power-constrained than the sensor nodes. The base-station transmits information and power to sensor nodes wirelessly via pencil beams created using an antenna array at the base-station. To overcome the range limitation of the MMID system [1], backscatter communication is not used in the PREMISS system. Instead, the sensor nodes first store the wireless received energy for a significantly longer period of time and use this energy only during a very short burst to transmit the information back to the base-station. Like this, the sensor nodes can transmit at a much higher power level than what is received from the base-station, thus can cover a larger range. In this chapter, we present a top-level system study of the PREMISS system to evaluate the system performance.

3.2 System Description

The general PREMISS system architecture is illustrated in Fig. 3.1. The system consists of monolithic integrated sensor nodes and a base-station. In the PREMISS system, in a single chip, a sensor node is integrated with the functions of wireless

© Springer International Publishing AG 2018   13
H. Gao et al., *Batteryless mm-Wave Wireless Sensors*, Analog Circuits
and Signal Processing, https://doi.org/10.1007/978-3-319-72980-0_3

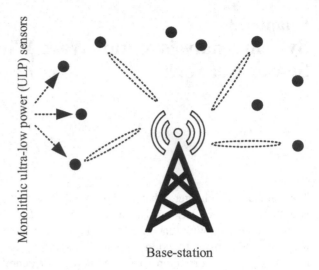

Fig. 3.1 General PREMISS system architecture

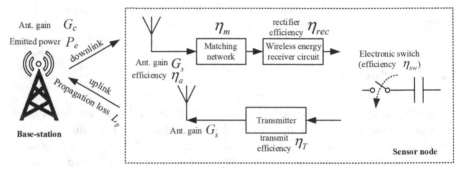

Fig. 3.2 Block diagram of the PREMISS system

power receiving, wireless sensing, and wireless data transfer. The base-station is a central controller for the wireless sensor nodes. At mm-wave frequencies, the base-station, with phased-array system architecture, can focus the power transfer to the location where sensor nodes are present by using beam-forming. The base-station transmits energy and information to the sensor nodes wirelessly, and receives the information transmitted back by the sensor nodes wirelessly. A typical sensing cycle consists of the two steps, as shown in Fig. 3.2:

- Downlink
  The base-station transfers the energy and commands wirelessly to the sensor nodes to wake up the sensor nodes and initiates the sensing.
- Uplink
  After the downlink operation, the sensor nodes start sensing the surrounding environment, and transfer the required information back to the base-station.

3.3 Link Budget Calculation

In this section, a link budget calculation for the PREMISS system is provided. In the system budget calculation among others, the on-chip antenna efficiency, the antenna matching, and the wireless power rectification efficiency are taken into consideration.

3.3.1 Downlink

As shown in Fig. 3.2, the base-station transmits with an emitted power P_e through the antenna array with array gain of G_c. The transmitted wireless signal is received by the on-chip antenna of the sensor nodes with the path loss, L_p. The on-chip antenna of the sensor nodes has the gain G_s and efficiency η_a. A matching network with an efficiency η_m is implemented between the antenna output and the input of the wireless on-chip power receiver module. The wireless on-chip power receiver has the efficiency of η_{rec}. The generated power is stored in the energy storage capacitor. There is a power switch connecting the energy storage capacitor and other active components of the sensor node, achieving the turn on and off function of the sensor nodes. The efficiency of this power switch is η_{sw}. Using these system parameters, the received power at the energy storage capacitor can be described as

$$P_S = \frac{P_e G_c G_s}{L_p} \eta_a \eta_m \eta_{sw} \eta_{rec} \tag{3.1}$$

and the energy stored in the capacitor, E_s, can be expressed in the first order with the charging time t_s

$$E_s = P_s t_s \tag{3.2}$$

The considerations and challenges for the parameters in (3.1) are discussed separately.

- Antenna array gain of the base-station, G_c:
  At 60 GHz, the wavelength in air is 5 mm. A phased array is used to increase the antenna gain. An antenna array with 1000 elements is assumed, which could provide extra antenna gain of 30 dBi.
- Free space loss, L_p:
  The free space loss can be expressed by Friis equation [2],

$$L_P = \left(\frac{4\pi d}{\lambda} \right)^2 = \left(\frac{4\pi df}{c_o} \right)^2 \tag{3.3}$$

where d is the distance between the base-station and the sensor node, λ and f are the wavelength and center frequency of the wireless signal, respectively. c_o is the speed of light, 3×10^8 m/s.

- Antenna gain at the sensor nodes, G_s:

The antenna gain could be expressed with its effective area A_{eff} [3] as

$$G = 4\pi \frac{A_{eff}}{\lambda^2} \tag{3.4}$$

For antenna sizes comparable to the wavelength of the signal or larger, A_{eff} is close to the actual antenna size. But if it is smaller than the wavelength, the maximum effective area can be calculated as

$$A_{eff} = D \frac{\lambda^2}{4\pi} \tag{3.5}$$

where D is the directivity of the antenna. The on-chip monopole antenna is $1/4$ of a wavelength, $\lambda/4$, thus the antenna gain can be described as

$$G_s = 4\pi \frac{A_{eff}}{\lambda^2} = \frac{4\pi}{\lambda^2} D \frac{\lambda^2}{4\pi} = D \tag{3.6}$$

- Efficiency of the on-chip antenna, η_a:

At mm-wave frequency, the silicon is not ideal for implementing on-chip antennas because of its low resistivity. The low resistivity of the silicon could cause a significant part of the radiated energy to be absorbed by the silicon substrate. This could be improved by using a high resistivity silicon substrate. In this chapter, we assume the on-chip antenna could reach 60% efficiency based on the reported work of [4].

- Efficiency of the matching network, η_m:

The matching network is used to match the antenna output impedance Z_A to the input impedance of the next stage, Z_L. Conjugate matching is used to maximize the power transfer, $Z_A^* = Z_L$. On-chip rectification is achieved through the rectifier, and its input impedance is a dynamic impedance, which changes with the input power and output DC voltage. This dynamic change will result in a mismatch between the antenna and the rectifier.

- Efficiency of rectification, η_{rec}:

The efficiency of the rectification depends on the input voltage swing. The input voltage swing can be transferred to the input power through the input impedance. In [5], a 7% efficiency is reported with the 5 dBm input power at 71 GHz.

- Efficiency of the power switch, η_{sw}:

In deep-sub micro CMOS technology, the switch efficiency can reach 90% [6].

Based on (3.2), the energy stored in the sensor node is plotted in Fig. 3.3, assuming P_e is 10 dBm, G_c is 30 dBi, G_s is 0 dBi, η_a is 0.6, η_m is 0.7, η_{rec} is 0.07, η_{sw} is 0.9, and the charging time t_s is 10 ms. From Fig. 3.3, it is shown that sensor nodes

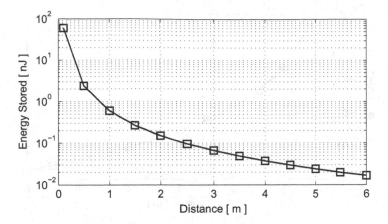

Fig. 3.3 Energy stored in the sensor nodes according to (3.2), assuming P_e is 20 dBm, G_c is 20 dBi, G_s is 0 dBi, η_a is 0.6, η_m is 0.7, η_{rec} is 0.07, η_{sw} is 0.9, and charging time t_s is 10 ms

could pick up 0.6 nJ at 1 m distance, and 0.024 nJ at 5 m distance. The PREMISS system targets sensor nodes for indoor applications, and 5 m is a typical distance for the indoor communication. In this system evaluation, the charging time of 10 ms is due to the on-chip capacitor value. During the 10 ms, a 2 nF on-chip capacitor could be charged to above 1 V. In other cases, this time could be longer with a larger capacitor in order to store more energy.

3.3.2 Uplink

As shown in Fig. 3.2, in the uplink, sensor nodes transmit the information back to the base-station by consuming the energy stored in the energy storage capacitor. Because the message that is transferred back to the base-station is very short, it can be transmitted within a short burst due to the high data rate in the 60 GHz band. In this way, the transmission power of the sensor node can be expressed as

$$P_T = \frac{E_s}{t_d} \tag{3.7}$$

where t_d is the burst duration. This transmission power can be larger than the received power, P_s, at the sensor nodes. For the uplink, the transmitter of the sensor node has an efficiency of η_T. The received signal power at the base-station, P_c can be described as

$$P_c = \frac{E_s}{t_d} \frac{G_s G_c}{L_p} \eta_T \tag{3.8}$$

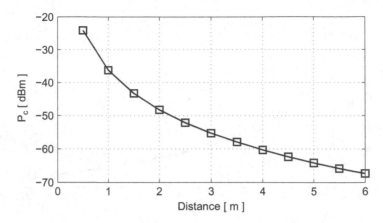

Fig. 3.4 Received power at the base-station according to (3.9), assuming uplink burst duration $t_d = 10$ ns, and transmission efficiency $\eta_T = 25\%$

where G_s, G_c, and L_P are the sensor node antenna gain, the base-station antenna gain, and the path loss. Combining (3.1) and (3.8), the received signal power at the base-station can be described with the system parameters as

$$P_c = \frac{t_s}{t_d} \frac{P_e G_s^2 G_c^2}{L_p^2} \eta_a \eta_m \eta_{sw} \eta_{rec} \eta_T \tag{3.9}$$

The received signal power at the base-station is plotted in Fig. 3.4 by using (3.9), assuming uplink burst duration $t_d = 10$ ns, and transmission efficiency $\eta_T = 25\%$.

The signal to noise ratio, SNR, at the base-station can be expressed as

$$SNR_c = \frac{P_c}{P_n} = \frac{P_c}{FW} \tag{3.10}$$

where $F = -174$ dBm/Hz is the power density of the thermal noise, and W is the bandwidth. Figure 3.5 shows the SNR can be achieved at the base-station for different distances between the sensor nodes and the base-station, assuming a bandwidth $W = 2$ GHz. It can be seen that for a range of 5 m, a decent SNR of 17 dB can still be achieved, which should be sufficient for the base-station to decode the sensor message successfully. This assumption is based on the first order estimation of the rectifier efficiency. With the input power decrease, the rectifier efficiency will decrease, thus the communication distance will be decreased.

It should be noted that in this link budget calculation for the PREMISS system, the energy required for sensing is not taken into consideration. The reason is that in the PREMISS system, the sensing is achieved through detecting temperature introduced frequency drift of the voltage controller oscillator (VCO) in the sensor nodes which is related to temperature variations. By exploring advanced algorithms at the base-station where more complexity and power consumption is affordable, the sensor node temperature can be extracted by measuring this frequency variation.

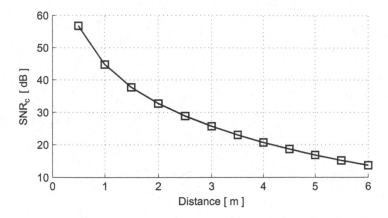

Fig. 3.5 SNR achieved at the base-station for the PREMISS system according to (3.10), assuming a bandwidth $W = 2\,\text{GHz}$

3.4 Conclusion

To overcome the size, cost, and robustness limitations of state-of-the-art sensor nodes, wireless power transfer is proposed in this chapter. A system study is performed with the considerations of the on-chip antenna, matching and components efficiencies. The downlink and uplink link budget is also provided in this chapter. However, in the system implementation, some factors including the Rx and Tx power consumption will increase the system power consumption and limit the system communication distance. Those effects will be discussed in Chap. 7.

References

1. A. Muller, D. Neculoiu, P. Pursula, T. Vaha-Heikkila, F. Giacomozzi, J. Tuovinen, Hybrid integrated micromachined receiver for 77 GHz millimeter wave identification systems, in *European Microwave Conference, 2007*, pp. 1034–1037 (2007)
2. H. Friis, A note on a simple transmission formula. Proc. IRE **34**(5), 254–256 (1946)
3. K.T. McDonald, Power received by a small antenna, Ph.D Dissertation, Princeton University, Princeton, Tech (2009)
4. U. Johannsen, A. Smolders, R. Mahmoudi, J. Akkermans, Substrate loss reduction in antenna-on-chip design, in *IEEE Antennas and Propagation Society International Symposium, 2009*, pp. 1–4 (2009)
5. H. Gao, M. Matters-Kammerer, D. Milosevic, A. van Roermund, P. Baltus, A 62 GHz inductor-peaked rectifier with 7% efficiency, in *2013 IEEE Radio Frequency Integrated Circuits Symposium (RFIC)*, June 2013, pp. 189–192
6. L. Chang, R. Montoye, B. Ji, A. Weger, K. Stawiasz, R. Dennard, A fully-integrated switched-capacitor 2:1 voltage converter with regulation capability and 90% efficiency at 2.3 A/mm², in *2010 IEEE Symposium on VLSI Circuits (VLSIC)*, pp. 55–56 (2010)

Chapter 4
Rectifier Analysis

Abstract In monolithic sensor networks, the rectifier is used as the on-chip wireless power receiver. In this chapter, the analysis and modeling of the rectifier are presented. Based on this analysis, a design flow for high efficiency rectifier development is presented.

4.1 Introduction

A rectifier is an electrical device converting an AC signal into a DC signal. This process is named rectification. Rectifiers are widely used in the role of generating current. A basic rectifier is composed of a device which forces the current flow into one direction. As the process of the rectification generates pulses of current, a filter is required for a steady and constant DC current through the load. Before the development of silicon semiconductor rectifiers, thermionic diodes and copper oxide or selenium-based rectifier stacks were used. Before the introduction of the semiconductor electronics, vacuum tube rectifiers were widely used. With the improvement of silicon technology, more compact devices are available for generating one directional current. Especially with the development of deep sub-micron technology, fast switching could be easily achieved by transistor based diodes or Schottky diodes.

In this chapter, the basic rectifier structure is introduced. Its key parameters related to wireless power transfer are presented, and the modeling of the rectifier is provided.

4.2 Basic Rectifier Structure

The basic rectifier structure is either a half-wave rectifier or full-wave rectifier, as shown in Fig. 4.1. The difference between them is apparent from the name. The half-wave rectifier only does the rectification either on the positive or negative half of the AC wave, while the other half is blocked. On the other hand, the rectification of

H. Gao et al., *Batteryless mm-Wave Wireless Sensors*, Analog Circuits and Signal Processing, https://doi.org/10.1007/978-3-319-72980-0_4

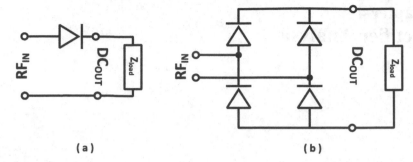

Fig. 4.1 Basic half-bridge and full-bridge rectifier. (**a**) Half-bridge rectifier. (**b**) Full-bridge rectifier

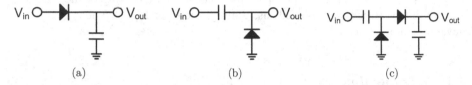

Fig. 4.2 Rectifier building blocks. (**a**) Basic rectifier. (**b**) Diode clamp. (**c**) Voltage doubler

the full-wave rectifier is achieved during the whole period of the input waveform. It converts both positive and negative phases of the input waveform to a unidirectional load current. Compared to the half-wave counterpart, the full-wave rectifier can provide a higher average output voltage with the assumption of the same input signal voltage amplitude. The half wave-form rectification will produce more ripple than the full-wave rectification, therefore a better filter to eliminate the harmonics from the output is required.

A voltage doubler is composed of two basic building blocks, the rectifier circuit and the clamp circuit [1]. The rectifier, shown in Fig. 4.2a, is used to generate a DC voltage from the input RF signal. When an input voltage is applied to the input of the circuit at the positive phase, the diode is on and the storage capacitor charges to the peak value of input voltage. If the capacitor is ideal, which means no resistor is connected in parallel with the output capacitor, the voltage V_{out} is kept constant. However, the leakage current of the capacitor induces an output voltage drop. If V_{in} is a sinusoidal wave, the capacitor will charge every time until it reaches its peak value. Thus, the mean voltage $\overline{V_{out}}$ is slightly smaller than the peak amplitude of V_{in}. Taking the threshold voltage of the diode into consideration the output voltage level will be further reduced. A clamping circuit, shown in Fig. 4.2b, is used to establish a DC reference for the output voltage by using a diode clamp. By conducting when the voltage at the output terminal of the capacitor V_{out} is negative, this circuit builds up an average charge on the terminal, which is sufficient to prevent the output from going negative. Positive charge on this terminal is effectively trapped. If all elements are ideal, the residual negative voltage ΔV_r is null and V_{out} is exactly equal to

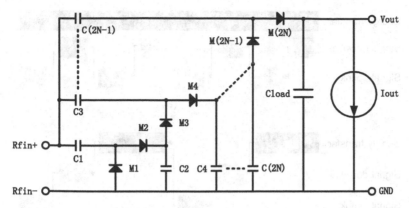

Fig. 4.3 Multi-stage rectifier

$V_{\text{inpeak}} + V_{\text{in}}$. The voltage doubler, shown in Fig. 4.2c, is composed of a rectifier circuit and a clamping circuit. The voltage doubler outputs a DC voltage. In the ideal situation, V_{out} is twice the amplitude of V_{in}.

By cascading a number of voltage doubler stages, the output voltage of the rectifier can be increased as compared to the peak AC input voltage. The structure was first proposed by Dickson [2], Fig. 4.3. The number of stages is limited by the current handing capacity, because each stage introduces extra loss which is generated from non-linear components.

4.3 Rectifier Performance Parameters

The performance of a rectifier is crucial for a wireless power transfer system. Its efficiency and sensitivity influences the system performance. In this section, two classes of wireless power transfer systems are introduced, one for high input power level and one for low input power level. The rectifier performance parameters related to those two classes of systems are provided.

4.3.1 General Wireless Power System Architecture

In wireless sensor networks with a wireless power transfer function, the energy is transferred wirelessly from the base-station to the sensor nodes. In principle, there are two types of the wireless powered systems, based on the sequence of the energy and data transfer. The first is the duplex system architecture and the second is the sequential system architecture. In the duplex architecture, energy is wirelessly received and consumed at the same time. In the sequential architecture, energy is temporarily stored, then consumed. The illustration is shown in Fig. 4.4.

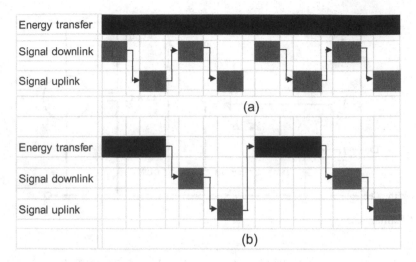

Fig. 4.4 Illustration of a duplex system and a sequential system. (**a**) Duplex system architecture. (**b**) Sequential system architecture

In a duplex system architecture, Fig. 4.4a, the power is wireless transferred from a base-station to the sensor nodes. At the same time, the power is consumed immediately by the sensor node. An RFID system is typically a duplex system architecture. In such a system, an RFID tag contains only a printed antenna and a chip, without batteries. A reader sends the power to the RFID tag, which rectifies the incident RF signal and generates the power supply for the tag system. The active part uses that energy and transfers the information with the backscatter method during the same time that the reader transfers the power.

In a sequential system architecture, Fig. 4.4b, there are separate time slots for wireless power transfer and wireless data transfer. The sensor node first stores the wirelessly received energy in an energy storage unit. Later on, the active components, which are inside the sensor node, consume this energy. In order to re-charge the energy storage unit, the RF power is transferred to the sensor nodes during the idle period of the active component. In this case, the stronger RF power signal will not block the RF signal for the wireless data transfer. Compared to the duplex system architecture, the sequential architecture has the possibility to provide more instant energy than the duplex architecture. The typical application for this system architecture is an ultra-low power duty-cycled sensor system. The sensor node will wake up periodically, and the average power consumption is low.

4.3.2 Rectifier Performance Parameters

Rectifiers are used in wireless power transfer systems in the wireless power receiver. For the rectifier, the most important parameters that will influence the system specification are sensitivity and efficiency.

The sensitivity is a measure for the relationship between the input power and the output voltage of the rectifier. In the sensor nodes of the wireless power transfer system, the rectifier has a function of rectification and provides the output voltage. The output voltage needs to be high enough to power up the active structures. In deep sub-micron technology such as 40 or 65 nm CMOS technology, a reasonable requirement for the power supply is 1 V. The sensitivity is defined as the input power level, which can provide the 1 V output DC voltage.

The efficiency of the rectifier (η_{rec}) is defined as the ratio between the output power and input power, which could be expressed as

$$\eta_{rec} = P_{out}/P_{in} \qquad (4.1)$$

For a fixed rectifier structure, there will be an optimum load current which could provide the best efficiency.

The sensitivity and efficiency requirements are different between the duplex system and sequential system architecture. In a duplex system, the rectifier needs to have higher efficiency to provide continuous output power, because the wirelessly received power is consumed immediately by the sensor node. The sensitivity of the rectifier in a duplex system should be sufficient for the required communication distance. In the sequential system, the rectifier efficiency can be lower because the energy will be stored temporarily inside the energy storage unit and this energy could be increased with longer transfer time. The efficiency of the rectifier in a sequential system should be large enough to compensate the system idle power consumption, mainly leakage current.

4.4 Rectifier Analysis and Modeling

Sensitivity and efficiency of the rectifier are the two most important performance parameters. The diode in the rectifier can be implemented by diode-connected NMOS transistor in CMOS technology [3]. When the input power is low, the transistor is operated close to the subthreshold region [4]. However, with increasing input power, the transistors are changing from the subthreshold region to the strong-inversion region [5]. This requires different modeling of the rectifier. In the subthreshold region, the transistor can be described by the diode equation. In the strong inversion region, it is necessary to use the square-law model of the transistor. Therefore, we apply two different models of the rectifiers: one is suitable for low input levels, when the transistor operates in the subthreshold region. The other one is suitable at higher input levels, when the transistor operates in the strong inversion region.

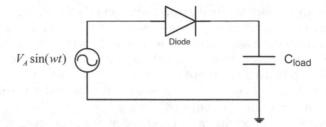

Fig. 4.5 Single-stage rectifier structure

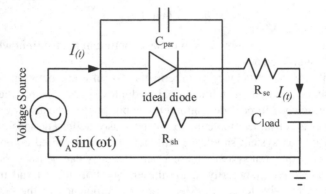

Fig. 4.6 Small-signal model of the single-stage rectifier

4.4.1   Modeling of Rectifier with Low Input Power

The RF-DC voltage rectifier is normally composed of a cascade of N single-stage rectifiers. If all the N stages use the same capacitors and diodes, the performance of the N-stage rectifier can be characterized with good approximation by the performance of a single-stage rectifier. As a result, the analysis starts with a single-stage rectifier as shown in Fig. 4.5. The input is a sinusoidal voltage source with amplitude V_A and frequency ω. The diode can be a junction diode, or a diode-connected MOS transistor and C_{load} is the storage capacitor.

Using a SPICE model for the diode [6], which is also a good approximation for a diode-connected MOS transistor in the subthreshold region [7], the equivalent small-signal (AC signal) model of the single-stage rectifier is depicted in Fig. 4.6. In the small-signal model, a practical diode is modeled with four components, namely an ideal diode, a series resistance R_{se} that models ohmic resistance from the diode, a parallel capacitance C_{par} that models the parasitic junction capacitances, and a shunt resistance R_{sh}. The shunt resistance is included in the SPICE model to aid convergence in the numerical calculations and has a default value of 10^{12} Ω [6]. It is thus not necessary in the analytical model that we are developing. However, in this model, R_{sh} is taken into account for the additional leakage current in the reverse bias region of a practical diode.

The current through an ideal diode due to a time varying voltage $V_d(t)$ is given by [6]

$$I_d(t) = I_s \left(\exp \left(\frac{V_d(t)}{NV_T} \right) - 1 \right), \tag{4.2}$$

where I_s is the saturation current, V_T is the thermal voltage, and N is the emission coefficient. The thermal voltage is given by $V_T = \frac{kT}{q} \approx 26\,\text{mV}$ at room temperature ($T \approx 300\,\text{K}$) and N has a value close to 1. The parasitic capacitance has a voltage dependent capacitance value given by

$$C_{par}(t) = C_0 \left(1 - \frac{V_d(t)}{V_0} \right)^{-0.5}, \tag{4.3}$$

where C_0 is the zero bias junction capacitance and V_0 is the contact potential having a typical value of 0.6–0.8 V. Assume the C_{load} is very large. The current through C_{par} can thus be written as

$$I_c(t) = \frac{d(C_{par}(t)V_d(t))}{dt} = C_{par}(t) \frac{\partial V_d(t)}{\partial t} + V_d(t) \frac{\partial C_{par}(t)}{\partial t}. \tag{4.4}$$

The current through R_{sh} is given by

$$I_R(t) = \frac{V_d(t)}{R_{sh}}, \tag{4.5}$$

and the total current equals $I(t) = I_d(t) + I_c(t) + I_R(t)$. The net electric charge accumulated on the storage capacitance in a charging cycle is thus given by

$$\Delta Q = \int_{t=0}^{T} I_d(t)dt + \int_{t=0}^{T} I_c(t)dt + \int_{t=0}^{T} I_R(t)dt, \tag{4.6}$$

where $T = 1/(\omega/(2\pi))$ is the period of the RF charging signal. For the case where we do not consider R_{sh}, $R_{sh} = \infty$ and $I_R(t) = 0$ in (4.6).

In the subsequent analysis, two key assumptions are proposed. First, the storage capacitance is large enough such that the voltage across the capacitance V_c in one charging cycle T is constant. Second, the voltage drop on R_{se} is negligible and $V_d(t) = V_A \sin(\omega t) - V_c$, where V_c is the voltage across C_{load}. When the circuit operates in a steady state and the load impedance is high, the output current is small. Therefore, the voltage drop on R_{se} is negligible.

4.4.1.1 Equilibrium Voltage

Using (4.2), we get

$$\int_0^T I_d(t)dt = \int_0^T I_s \left(\exp\left(\frac{V_A \sin(\omega t) - V_c}{NV_T} \right) - 1 \right) dt$$

$$= I_s T \left(\exp\left(-\frac{V_c}{NV_T} \right) \mathcal{I}_0 \left(\frac{V_A}{NV_T} \right) - 1 \right), \tag{4.7}$$

The appearance of the modified Bessel function $\mathcal{I}_0(x)$ of the first kind and order zero was recognized by Harrison [8] in his analysis of analog TV reception. Because of symmetry properties over one sinusoidal cycle in (4.4), it can be shown that $\int_{t=0}^T I_c(t)dt = 0$. In the steady state of the rectifier, the electric charge that enters C_{load} should be equal to the electric charge leaving it. Therefore, $\Delta Q = 0$ for one charging cycle. As a result, without considering R_{sh}, from (4.7), we have

$$\left(\exp\left(-\frac{\widetilde{V_c}}{NV_T} \right) \mathcal{I}_0 \left(\frac{V_A}{NV_T} \right) - 1 \right) = 0, \tag{4.8}$$

where $\widetilde{V_c}$ is the equilibrium voltage, and

$$\widetilde{V_c} = NV_T \log\left(\mathcal{I}_0 \left(\frac{V_A}{V_T} \right) \right). \tag{4.9}$$

Previous studies on scavenging rectifiers relied on numerical methods to evaluate V_c [7, 9]. The Bessel function $\mathcal{I}_0$ allows us to phrase V_c as the analytical expression (4.9) and also appears instrumental for further evaluation. Expression (4.9) is valid for weak RF signals that keep the diode in the subthreshold regime. In fact, a first order Taylor's series approximation of the Bessel function yields $V_c \approx \frac{V_A^2}{4NV_T}$ and confirms the highly non-linear behavior between the rectifier input voltage and the equilibrium voltage that poses design challenges.

To include R_{sh}, the electric charge that goes through it also needs to be taken into account. Using (4.5), we get

$$\int_{t=0}^T I_R(t)dt = -\frac{V_c}{R_{\text{sh}}}T. \tag{4.10}$$

Again, when $\Delta Q = 0$, we have

$$I_s \left(\exp\left(-\frac{\widetilde{V_c^R}}{NV_T} \right) \mathcal{I}_0 \left(\frac{V_A}{NV_T} \right) \right) - \left(I_s + \frac{V_c^R}{R_{\text{sh}}} \right) = 0, \tag{4.11}$$

and the equilibrium voltage $\widetilde{V_c^R}$, considering R_{sh}, can be obtained numerically. From (4.11), $\widetilde{V_c^R}$ can be expressed as

$$\widetilde{V_c^R} = NV_T \log\left(\mathcal{I}_0\left(\frac{V_A}{NV_T}\right)\right) - NV_T \log\left(1 + \frac{\widetilde{V_c^R}}{I_s R_{sh}}\right). \qquad (4.12)$$

For $\frac{\widetilde{V_c^R}}{I_s R_{sh}} \ll 1$, using Taylor's series approximation, $\log\left(1 + \frac{\widetilde{V_c^R}}{I_s R_{sh}}\right)$ can be simplified as $\frac{\widetilde{V_c^R}}{I_s R_{sh}}$ and we can obtain the closed-form approximation for the equilibrium voltage

$$\widetilde{V_c^R} \approx \frac{NV_T \log\left(\mathcal{I}_0\left(\frac{V_A}{NV_T}\right)\right)}{1 + \frac{NV_T}{I_s R_{sh}}}. \qquad (4.13)$$

Comparing (4.13) with (4.9), it could be found that including R_{sh} results in a smaller equilibrium voltage. The effect of R_{sh} which is shown on the term $\frac{NV_T}{I_s R_{sh}}$ in the denominator, accounts for the additional leakage current through R_{sh} in the reversed-bias region.

4.4.1.2 Input Resistance

The input resistance is an important parameter for designing the matching network that maximizes power transfer from the antenna to the rectifier. Due to the non-linear current through the diode, the input resistance is no longer a constant [9]. In this section, we determine an average input resistance $\overline{R_{in}}$ related to the mean power that enters the rectifier in one charging cycle [9]

$$\overline{P_{in}} = \frac{1}{T}\int_0^T V_A \sin(\omega t) I(t) dt$$

$$= \frac{1}{T}\int_0^T V_A \sin(\omega t)(I_d(t) + I_c(t) + I_R(t)) dt. \qquad (4.14)$$

Using (4.2), we obtain

$$\frac{1}{T}\int_0^T V_A \sin(\omega t) I_d(t) dt = \frac{1}{T}\int_0^T V_A I_s \left(\exp\left(\frac{V_d(t)}{NV_T}\right) - 1\right)\sin(\omega t) dt$$

$$= V_A I_s \exp\left(-\frac{V_c}{NV_T}\right)\mathcal{I}_1\left(\frac{V_A}{NV_T}\right)$$

$$= V_A I_s \frac{\mathcal{I}_1\left(\frac{V_A}{NV_T}\right)}{\mathcal{I}_0\left(\frac{V_A}{NV_T}\right)}, \qquad (4.15)$$

where $\mathcal{I}_1(x)$ is the modified Bessel function of the first kind and order 1. Using (4.4), it can be shown that $\frac{1}{T}\int_0^T V_A \sin(\omega t)I_c(t)dt = 0$. As a result, without considering R_{sh}, we have

$$\overline{P_{in}} = V_A I_s \frac{\mathcal{I}_1\left(\frac{V_A}{NV_T}\right)}{\mathcal{I}_0\left(\frac{V_A}{NV_T}\right)}, \qquad (4.16)$$

and the average input resistance over a charging cycle becomes

$$\overline{R_{in}} = \frac{1}{2}\frac{V_A^2}{\overline{P_{in}}} = \frac{V_A}{2I_s}\frac{\mathcal{I}_1\left(\frac{V_A}{NV_T}\right)}{\mathcal{I}_0\left(\frac{V_A}{NV_T}\right)}. \qquad (4.17)$$

When considering R_{sh}, using (4.5), we get

$$\frac{1}{T}\int_0^T V_A \sin(\omega t)I_R(t)dt = \frac{V_A^2}{2R_{sh}}, \qquad (4.18)$$

and

$$\overline{P_{in}} = V_A I_s \exp\left(-\frac{V_c^R}{NV_T}\right)\mathcal{I}_1\left(\frac{V_A}{NV_T}\right) + \frac{V_A^2}{2R_{sh}}. \qquad (4.19)$$

The average input resistance in this case is equal to

$$\overline{R_{in}^R} = \left(\frac{2I_s \exp\left(-\frac{\widetilde{V_c^R}}{NV_T}\right)\mathcal{I}_1\left(\frac{V_A}{NV_T}\right)}{V_A} + \frac{1}{R_{sh}}\right)^{-1}. \qquad (4.20)$$

To get more insight from (4.20), we assume $\widetilde{V_c^R} \approx \widetilde{V_c}$, then

$$\overline{R_{in}^R} \approx \left(\frac{1}{\overline{R_{in}}} + \frac{1}{R_{sh}}\right)^{-1}. \qquad (4.21)$$

and this is the equivalent resistant of a parallel-connected $\overline{R_{in}}$ as in (4.17) and R_{sh}.

4.4.1.3 Charging of the Storage Capacitor

To be able to describe the charging of C_{load} over time, we propose a quasi-static charging model. We assume that the voltage on C_{load} stays constant within T, while the voltage increment from nT to $(n+1)T$ is given by

$$V((n+1)T) - V(nT) = \frac{\int_{t=nT}^{(n+1)T} I(t)dt}{C_{\text{load}}}$$

$$= \frac{T}{C_{\text{load}}} \left(I_s \left(\exp\left(-\frac{V(nT)}{NV_T}\right) \mathcal{I}_0 \left(\frac{V_A}{NV_T}\right) - 1 \right) - \frac{V(nT)}{R_{\text{sh}}} \right), \quad (4.22)$$

and with this, we can determine the voltage on the capacitor in a given time period. Similarly, we can obtain that the energy accumulated during one charging cycle is given by

$$E((n+1)T) - E(nT) = \frac{1}{2} C_{\text{load}} \left(V^2((n+1)T) - V^2(nT) \right), \quad (4.23)$$

thus the power saved can be derived from (4.23) as

$$P(nT) = \frac{1}{2T} C_{\text{load}} \left(V^2((n+1)T) - V^2(nT) \right). \quad (4.24)$$

4.4.1.4 Comparison with Circuit Simulation Results

In this part, the performance of the one-stage rectifier calculated using our model is compared with that obtained using the Cadence Virtuoso Spectre circuit simulator. The diode is implemented as diode-connected transistor in 65 nm CMOS technology with a channel length of 60 nm, a channel width of 2 μm and has three fingers. Firstly, the DC performance of the diode-connected transistor is compared. Figure 4.7 shows the DC current through the diode as a function of the DC voltage

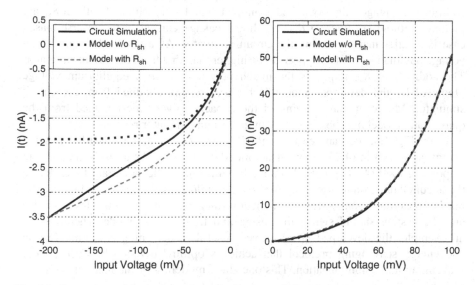

Fig. 4.7 Comparison of the model and circuit simulator—DC performance

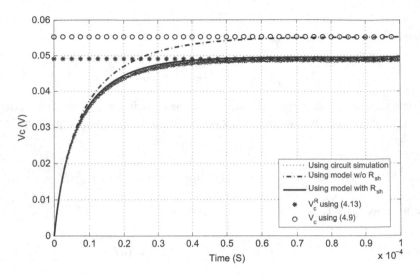

Fig. 4.8 Comparison of the capacitor charging performance using the analytical model and the circuit simulator

for the reverse biased and forward biased (subthreshold) regions obtained using the model and the circuit simulator. In the forward biased region this model agrees well with circuit simulator. In the reverse biased region, without using the shunt resistance, the DC (leakage) current obtained using the model is too large compared to that obtained using the circuit simulator. With the shunt resistance, the current becomes similar.

Figure 4.8 shows the comparison of the voltage build-up on C_{load} obtained using the quasi-static charging model (4.22) and using the Cadence circuit simulator. Two equilibrium voltage values obtained using (4.13) and (4.9) with and without R_{sh} are plotted. Without considering R_{sh}, the diode leakage current in the reverse-biased case is smaller than that using the circuit simulator. As a result, the equilibrium voltage obtained from (4.9) appears to be higher than that from the circuit simulator. The addition of R_{sh} improves the match between the final equilibrium voltage predicted by the developed model (4.13) and results obtained from the circuit simulator. Moreover, the charging of the capacitor over time obtained from the quasi-static model matches closely with the circuit simulator results.

From Fig. 4.8, we can see that initially, the voltage grows linearly, thus the energy as shown in (4.23) grows quadratically with time. As a result, the amount of power stored (4.24) grows also linearly with time. After a certain time, due to the accumulated voltage on C_{load}, the voltage build-up slows down and eventually reaches the equilibrium. After this moment further scavenging is much less effective and the system should switch to sensing and transmission mode. Therefore, as an example, the developed model is very useful in designing a communication and energy scavenging protocol that achieves optimal division between energy scavenging and communication. This one-stage modeling of the rectifier with low input power can be extended into the N-stage rectifier. Because of the research limitation, this N-stage extension is recommended for future direction.

4.4.2 Modeling of Rectifier with High Input Power

In this paragraph a model for the charging of the hold capacitor through an N-stage rectifier will be derived for input RF signals large enough to operate in the strong inversion region.

For N-stage multi-stage rectifiers, the average output voltage can be expressed as

$$V_{\text{out}} = N \times V_{\text{boost}} \tag{4.25}$$

where V_{boost} is the incremental voltage of each stage, the analysis of the multi-stage rectifier can be simplified to analysis the V_{boost} from single stage. V_{boost} can be expressed in (4.26) [7].

$$V_{\text{boost}} = V_a' - V_{\text{th}} - \left(\frac{15\pi}{8} \frac{I_{\text{oeff}}' \sqrt{2V_a'}}{\mu_n C_{\text{ox}} \frac{W}{L}} \right)^{\frac{2}{5}} \tag{4.26}$$

where V_{th} is the transistor threshold voltage, μ_n is the electron mobility, C_{ox} is the oxide capacitance, V_a' is the effective input rectifier voltage swing and can be expressed as (4.27), I_{oeff}' is the effective output current of the each stage and can be expressed in (4.28).

$$V_a' = \frac{C}{C + C_{\text{par}}} V_a \tag{4.27}$$

$$I_{\text{oeff}}' = I_o + \frac{I_{\text{so}}}{\pi} \frac{W}{L} \left(1 - e^{-\frac{V_a'}{V_T}} \right) \left(1 + \lambda_{\text{sub}} V_a' \right) \tag{4.28}$$

where I_{so} is the leakage current, V_T is the thermal voltage, λ_{sub} is the subthreshold region channel-length modulation parameter, C_{par} is the input referred capacitance and can be expressed as

$$C_{\text{par}} = \alpha C + 2C_{\text{mos}} \tag{4.29}$$

where α is a proportionality constant relating the upper-plate parasitic capacitance to the storage capacitance. C_{mos} is the overlap and junction capacitance of two adjacent transistors and can be expressed in (4.30).

$$C_{\text{mos}} = C_{\text{gs0}} + C_j WE + 2C_{\text{jsw}}(W + E) \tag{4.30}$$

where C_{gs0} is the overlap capacitance per unit width, C_j is the junction capacitance per unit area, and C_{jsw} is the side-wall junction capacitance per unit perimeter.

The total power dissipation of the one stage rectifier is composed of power dissipation from the MOS transistor when working in the strong inversion region and the power dissipation from the leakage current when the input voltage swing is smaller than the transistor threshold voltage. It can be expressed in (4.31), where the power dissipation of the rectifier with MOS transistors in the strong inversion region is given by (4.32) and the power dissipation due to the leakage current is expressed in (4.33) [7].

$$P_{loss} = P_{inv} + P_{leak} \tag{4.31}$$

$$P_{inv} = \frac{6}{7} I_{oeff} \left(V_a + \frac{1}{6} V_{thn} - V_{boost} \right) \tag{4.32}$$

$$P_{leak} = I_{so} \frac{W}{L} \left[\frac{\overline{V_o}}{2} + \frac{V_a}{\pi} + \lambda_{sub} \left(\frac{\overline{V_o}^2}{2} + \frac{V_a^2}{\pi} + \frac{2\overline{V_o}V_a}{\pi} \right) \right] \tag{4.33}$$

where $\overline{V_o}$ is the output voltage averaged over time.

Equation (4.31) can be re-written by combining (4.32) and (4.33),

$$P_{loss} = \frac{6}{7} I_{oeff} \left(V_a + \frac{1}{6} V_{thn} - V_{boost} \right)$$
$$+ I_{so} \frac{W}{L} \left[\frac{\overline{V_{boost}}}{2} \frac{V_a}{\pi} + \lambda_{sub} \left(\frac{\overline{V_{boost}}^2}{2} + \frac{V_a^2}{\pi} + \frac{2\overline{V_{boost}}V_a}{\pi} \right) \right] \tag{4.34}$$

The power dissipation for a multi-stage rectifier can be expressed by integrating the individual power dissipation of each stage, $P_{lossN} = \sum_{n=1}^{N} P_{loss|n} \approx N \times P_{loss}$. So the efficiency of a multi-stage rectifier can be expressed as

$$\eta = \frac{V_{out} \times I_{load}}{P_{lossN} + V_{out} \times I_{load}}$$
$$= \frac{N \times V_{boost} \times I_{load}}{N \times P_{loss} + N \times V_{boost} \times I_{load}}$$
$$= \frac{V_{boost} \times I_{load}}{P_{loss} + V_{boost} \times I_{load}} \tag{4.35}$$

From (4.35), it is noticed that the multi-stage rectifier efficiency is independent of the number of stages and that it is equal to the single-stage rectifier efficiency. Thus, optimization of a single stage leads to the optimization of the multi-stage rectifier.

After this analysis of the rectifier, there are four important observations that can be brought into the design process.

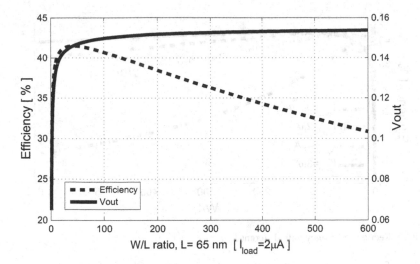

Fig. 4.9 Output voltage and efficiency with different W/L ratio

4.4.2.1 Choice of W/L

The first important observation is based on (4.26) and (4.35): the optimization of the width and length ratio for maximum efficiency does not correspond to the optimum for the maximum output voltage. Assume a one stage rectifier with fixed intermediate storage capacitor value, then the voltage drop across the diode-connected transistor should be small in order to have a higher output voltage, which requires a larger transistor with low drain-source resistance, r_{ds}. However, the bigger the transistor, the larger the parasitic transistor capacitors and consequently also the larger the leakage current lowering the efficiency. On the other hand, keeping the transistor W/L small will make the charge transfer incomplete within the available time set by the RF input, which will lower the efficiency. Figure 4.9 shows that optimum efficiency does not correspond to maximum output voltage.

4.4.2.2 Maximum Efficiency

The second observation comes from the expression for the efficiency of the rectifier in (4.35). With different requirements of output current, I_{load}, the rectifier with different width and length ratios can reach the same maximum efficiency, as shown in Fig. 4.10. In Fig. 4.10, with the output current varying from 0.5 to $8\,\mu A$, the maximum efficiency is the same after properly adapting the width and length ratio.

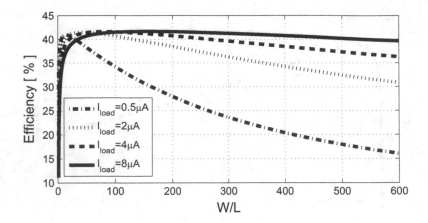

Fig. 4.10 Rectifier efficiency with different I_{load}

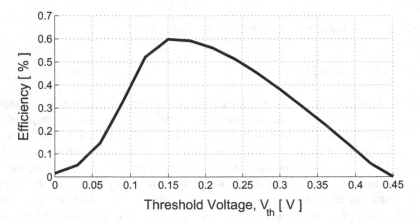

Fig. 4.11 Rectifier efficiency with V_{th}

4.4.2.3   Relation Between Efficiency and Threshold Voltage

The third observation is that there is a relationship between the MOS transistor threshold voltage and the rectifier efficiency as shown in Fig. 4.11. The lower the threshold voltage, the quicker the MOS diode is turned on, and the more saturation current it will allow to flow. However, at the same time with lower threshold voltage more leakage current is to be expected as well [10]. Thus, there should be an optimum threshold voltage for a given technology.

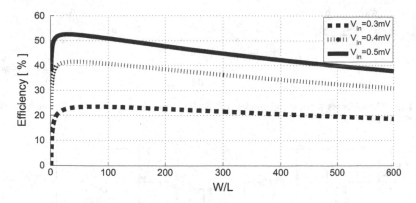

Fig. 4.12 Rectifier efficiency with different input voltage swings

4.4.2.4   Relation Between Efficiency and Input Voltage

The fourth observation is related to the influence of the input voltage swing on the efficiency (4.35). This is shown in Fig. 4.12. The input voltage has a big impact on the performance of the rectifier, as it directly controls the turning on and off of the MOS diode and the amount of current. The simulated efficiency for a given rectifier output voltage confirms the strong relation with the input voltage swing. In wireless sensor network, the input voltage swing can be increased with the same input power, e.g. by an impedance matching network or by a transformer.

4.5   Limitations of Rectifier Modeling and Challenges

4.5.1   Rectifier Modeling Limitation

In this chapter, the modeling of the rectifier is mainly illustrated for the low frequency range, and it can support the frequency up to 2 GHz. With frequency increasing to the mm-wave range, this modeling will not be accurate. The difference between the low frequency range and the mm-wave frequency range behavior is mainly due to the parasitic capacitors of the transistors. Figure 4.13a shows the cross-sectional view of a typical bulk NMOS transistor. Small-signal circuit models for the ON state and the OFF state of the NMOS transistor are shown in Fig. 4.13b and c. In the ON state, the on-resistance R_{on} determines the insertion loss at low frequencies. As the operating frequency increases, insertion loss will increase due to the increase of capacitive coupling of the signal through the ON-state parasitic capacitances C_{db}, C_{sb}, C_{gd}, C_{gs}, and C_{gb} to the substrate resistance R_B. In the OFF state, off-resistance R_{off} determines the isolation at low frequencies. R_{off} is typically more than one hundred $k\Omega$ leading to very good isolation. However, at high frequency, isolation becomes worse due to parasitic capacitances C_{db}, C_{sb}, C_{gd}, C_{gs}, and C_{gb} and the channel capacitance C_{off}.

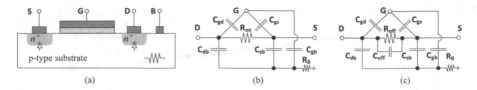

Fig. 4.13 NMOS cross section and model. (**a**) NMOS cross section. (**b**) NMOS on-state. (**c**) NMOS off-state

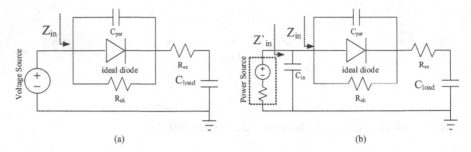

Fig. 4.14 The improvement of small-signal model of a single-stage rectifier. (**a**) Model of single-stage rectifier at low frequency. (**b**) Model of single-stage rectifier at mm-wave frequency

Based on the analysis of the differences between the low frequency model and the mm-wave frequency model, the modeling should take into account these parasitic capacitors. The model in Fig. 4.14a should be replaced by the model in Fig. 4.14b. First, the input voltage source should be replaced by a power source. Second, C_{in} should be included, thus $Z'_{in} = Z_{in}//C_{in}$. C_{in} is the parasitic capacitor to the ground at the input terminal, which is dominated by C_{gs} of the transistor and the Miller capacitor from C_{gd}. With increasing frequency, $|Z'_{in}|$ will decrease due to the influence of C_{in}, and this variation will change the input voltage swing. This effect could be taken into account by replacing the voltage source by power source. Because $|V_{in}| = \sqrt{P_{in}|Z'_{in}|}$, the input voltage swing will decrease with increasing frequency. At mm-wave frequencies, the current flow through C_{par} will provide a path from the input to the output, which would discharge the load capacitor directly. Thus, this current should be taken into account in the modeling as well.

4.5.2 mm-Wave Rectifier Challenges

The RF rectifier performance is changing with frequency. At 90 MHz and 900 MHz for the RFID applications, the input sensitivity of the rectifier can reach −20 dBm or even better [11], and the efficiency can reach up to 60% [12], while at the mm-wave frequencies, such as 40 GHz, the efficiency is only 1.4% [13].

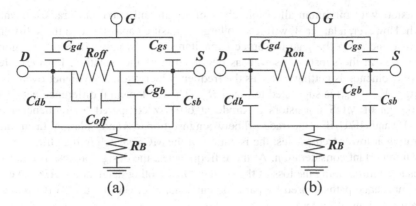

Fig. 4.15 NMOS transistor model. (**a**) Off state model. (**b**) On state model

4.5.2.1 Efficiency

Power loss in the rectifier limits its efficiency. The loss of the rectifier comes from two parts. The first one is through the transistor itself. The second is through the high frequency signal coupling to the output storage capacitor, bad isolation. The NMOS transistor model is shown in Fig. 4.15, where C_{gs}, C_{gd}, C_{db}, C_{gb}, and C_{sb} are the parasitic capacitors transistor. R_{on} is the on-resistance of the transistor, R_{off} is the off-resistance of the transistor. R_B is the substrate resistivity. The parasitic capacitors C_{db} and C_{sb} provide the signal path to the substrate (R_B). Because the substrate is a lossy material, this coupling will introduce power dissipation.

When the transistor is turned on, the loss at low frequencies is determined by the on-resistance of the NMOS channel, R_{on}. At mm-wave operating frequency, the loss will increase due to the capacitive coupling of the signal through the parasitic capacitances of the transistor to the substrate.

When the transistor is turned off, the isolation at low frequencies is determined by the off-resistance R_{off}. For higher frequencies, the isolation from input to output becomes worse due to the off-state parasitic capacitances and channel capacitance, therefore the storage capacitor, which is charged by the DC output voltage of the rectifier, is discharged by the signal fed through the parasitic capacitors. To increase the efficiency of the rectifier, the insertion loss must be minimized and the isolation maximized. In order to obtain minimum insertion loss, one needs to reduce the resistance R_{on}. The on-resistance can be expressed [14] as

$$R_{on} = \frac{1}{\mu C_{ox} \frac{W}{L} \left(V_{gs} - V_{th}\right)} \tag{4.36}$$

where μ is the carrier mobility, C_{ox} is the gate oxide capacitance per unit area, V_{th} is the transistor threshold voltage, V_{gs} is the gate-source voltage, W and L are the width and length of the transistor, respectively. That suggests the use of transistors with minimum allowable channel length and maximum transistor gate width. However, a larger W will lead to larger parasitic capacitance to the substrate and also increases the gate to source capacitance C_{gs}, which will result in more signal loss in the substrate as well as degradation of the isolation. Furthermore, this phenomenon is getting worse as the frequency increases. Also higher overdrive voltage ($V_{gs} - V_{th}$) is suggested to lower R_{on}. The fundamental difference between the design of MOS transistors as diode at 60 GHz compared to frequencies of 2.4 GHz and HF/UHF is the trade-off between insertion loss and isolation. In circuits operating at lower frequencies, the isolation of the MOS switch in the OFF state is not an important consideration. At these frequencies, the design process is entirely focused on minimizing the loss of the switch. On the other hand, at 60 GHz, several low-impedance paths caused by parasitic capacitances lead to a trade-off between isolation and insertion loss.

4.5.2.2   Sensitivity

The sensitivity of the rectifier is the input power required to provide 1 V DC output voltage. The rectifier is composed of the switch which is implemented by the CMOS transistor. The switch is sensitive to the input voltage swing. With fixed input power of the mm-wave signal, the input voltage swing is determined by the input impedance. When the transistor turns on or off is defined by the threshold voltage. Thus, the sensitivity is limited by both the input impedance and the transistor threshold voltage.

4.6   Conclusion

In this chapter, considerations of using an RF-DC rectifier in wirelessly powered systems have been presented. The transistors which are used in rectifiers to form the diodes are working in different regions depending on the input power level. Two models are built based on the two different cases. The first model is for an RF rectifier with low input power and this model can be used for rectifier sensitivity analysis in case of low input power levels. In this model, the key parameters are equilibrium voltage and input resistance and are described in terms of Bessel functions. An analytic expression for input power and input impedance in terms of higher order Bessel functions is derived. This leads to results for the dynamics of energy built up in RF wireless powered sensor nodes, its equilibrium value and the time constants of reaching this. The performance predicted by this model matches closely with that obtained using a circuit simulator. This model provides an important building block for system level performance analysis and optimization to address the increasing demand to extend the operation range and/or reduce cost and size of state-of-the-art self-powered wireless sensors. The second model is for an RF rectifier in the high input power regime and this model can be used for rectifier efficiency analysis in case of high input power levels. In this model, the key parameters such as input power, transistor sizing, and transistor threshold voltage are described in the spice-mode based equation. In the end of this chapter, the limitations of the rectifier modeling for mm-wave frequencies are pointed out and future recommendations are provided. Based on the analysis of the challenges at mm-wave frequencies, designs of mm-wave frequencies rectifiers are provided in the next chapter.

References

1. J.-P. Curty, M. Declercq, C. Dehollain, N. Joehl, *Design and Optimization of Passive UHF RFID System* (Springer, Boston, 2007)
2. J. Dickson, On-chip high-voltage generation in MNOS integrated circuits using an improved voltage multiplier technique. IEEE J. Solid-State Circuits **11**(3), 374–378 (1976)
3. J. Wang, Y. Jiang, J. Dijkhuis, G. Dolmans, H. Gao, P.G.M. Baltus, A 900 MHz RF energy harvesting system in 40 nm CMOS technology with efficiency peaking at 47% and higher than 30% over a 22dB wide input power range, in *43rd Proc. ESSCIRC 2005*, September 2017
4. Y. Wu, J.-P. Linnartz, H. Gao, M. Matters-Kammerer, P. Baltus, Modeling of RF energy scavenging for batteryless wireless sensors with low input power, in *2013 IEEE 24th International Symposium on Personal Indoor and Mobile Radio Communications (PIMRC)*, September 2013, pp. 527–531
5. H. Gao, P. Baltus, R. Mahmoudi, A. van Roermund, 2.4 GHz energy harvesting for wireless sensor network, in *2011 IEEE Topical Conference on Wireless Sensors and Sensor Networks (WiSNet)*, January 2011, pp. 57–60
6. W. Roehr (ed.), *Rectifier Applications Handbook: Reference Manual and Design Guide*, 2nd edn. (Semiconductor Components Industries, LLC, 2011)
7. J. Yi, W.-H. Ki, C.Y. Tsui, Analysis and design strategy of UHF micro-power CMOS rectifiers for micro-sensor and RFID applications. IEEE Trans. Circuits Syst. I Regul. Pap. **54**(1), 153–166 (2007)
8. R. Harrison, Full nonlinear analysis of detector circuits using Ritz-Galerkin theory, in *IEEE MTT-S International Microwave Symposium Digest, 1992*, vol. 1, June 1992, pp. 267–270
9. J.-P. Curty, N. Joehl, F. Krummenacher, C. Dehollaini, M. Declercq, A model for μ-power rectifier analysis and design. IEEE Trans. Circuits Syst. I Regul. Pap. **52**(12), 2771–2779 (2005)
10. K. Roy, S. Mukhopadhyay, H. Mahmoodi-Meimand, Leakage current mechanisms and leakage reduction techniques in deep-submicrometer CMOS circuits. Proc. IEEE **91**(2), 305–327 (2003)
11. L. Xia, J. Cheng, N. Glover, P. Chiang, 0.56 V, −20 dBm RF-powered, multi-node wireless body area network system-on-a-chip with harvesting-efficiency tracking loop. IEEE J. Solid-State Circuits **49**(6), 1345–1355 (2014)
12. A. Bakhtiar, M. Jalali, S. Mirabbasi, A high-efficiency CMOS rectifier for low-power RFID tags, in *2010 IEEE International Conference on RFID*, April 2010, pp. 83–88
13. S. Pellerano, J. Alvarado, Y. Palaskas, A mm-wave power-harvesting RFID tag in 90 nm CMOS. IEEE J. Solid-State Circuits **45**(8), 1627–1637 (2010)
14. B. Razavi, *RF Microelectronics*, 2nd edn. (Prentice Hall, Upper Saddle River, 2011)

Chapter 5
mm-Wave Rectifiers

Abstract In the previous chapters, the analysis of the rectifier for wireless power transfer is presented. Issues of the rectifiers working at the mm-wave frequency are analyzed. In this chapter, based on the discussion of the mm-wave rectifier, solutions are provided to increase the rectifier efficiency, including the inductor-peaking method, local threshold voltage modulation, and increased isolation by output filtering. At the end of this chapter, the implementations and measurement results of mm-wave rectifiers in 65 nm CMOS technology are provided.

5.1 Introduction

The most challenging issue for fully integrated sensor nodes is to provide enough power wirelessly on chip. In [1], a multi-stage rectifier based on the Dickson structure is implemented in 90 nm CMOS technology, and achieved 8% efficiency with 2 dBm input power with 1 V output voltage at 40 GHz. Reference [2] adapts off-chip body-bias technology to increase the sensitivity, and achieves 1 V output voltage with 2 dBm input power at 60 GHz. In this chapter, three methods are proposed to increase the efficiency. The method of inductor peaking is proposed to decrease the on-resistivity and increase the second order harmonic generation. The method of local-bias modulation is proposed, to decrease the threshold so as to further decrease the loss from the transistor. An output filter is used to increase the isolation from input to output so as to prevent the discharging current in order to increase efficiency. Three versions of the mm-wave rectifiers with measured results are provided. The summary and conclusion are presented at the end.

5.2 Methods to Improve the mm-Wave Rectifier Performance

The efficiency of the rectifier at the mm-wave frequencies is limited by three mechanisms. The first one is the loss in the transistor channel. The second one is bad isolation between input and output. The third is the RF signal coupling to

© Springer International Publishing AG 2018
H. Gao et al., *Batteryless mm-Wave Wireless Sensors*, Analog Circuits
and Signal Processing, https://doi.org/10.1007/978-3-319-72980-0_5

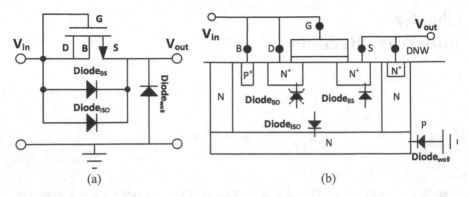

Fig. 5.1 (**a**) Schematic of body-drain connection and (**b**) cross-view of deep N-well technology

the lossy substrate through parasitic capacitors. For transistor loss, decreasing R_{on} by increasing the value of the overdrive voltage, $V_{gs} - V_{th}$, is a possible solution. That means, either increasing the value of V_{gs} or decreasing V_{th} could improve the efficiency. Also, the sensitivity of the rectifier can be increased by decreasing V_{th}. The local threshold voltage modulation and the inductor-peaking methods are used to achieve this. The isolation problem can be solved by using a filter at the DC output. However, the loss introduced by the parasitic capacitor to the substrate can only be improved by optimizing the parasitic capacitors by changing the layout of the MOS transistor or by increasing the substrate resistivity.

5.2.1   Threshold Voltage Modulation

The cross-view of a deep N-well NMOS transistor and its application in the rectifier are shown in Fig. 5.1. During the positive phase of the input voltage swing and the conducting phase of the diode-connected transistor, the loss is dominated by the R_{on} of the transistor. During the negative phase of the input voltage, the loss is due to the parasitic diode, Diode$_{BD}$ and parasitic capacitor, C_{GS}. In the negative phase, the parasitic diode, Diode$_{BD}$, will discharge the load capacitor. At mm-wave frequencies, the parasitic capacitor, C_{GS}, will provide a low-impedance path and thereby partly discharge the load capacitor. In the negative phase of the cycle, the leakage current of the transistor will also decrease the efficiency of the rectifier. In deep n-well technology, the bulk can be connected to the drain. In this way, Diode$_{BD}$ is short-circuited and will not influence the efficiency and sensitivity of the rectifier, as shown in Fig. 5.1b. The threshold voltage will be modulated by the input voltage [3]. In the positive phase of the input voltage swing, the introduced current will be larger than in the case of the normal bulk-source connection at the same input voltage level. In the negative phase of the input voltage swing, the leakage current will be smaller than normal connection. The proposed technique of

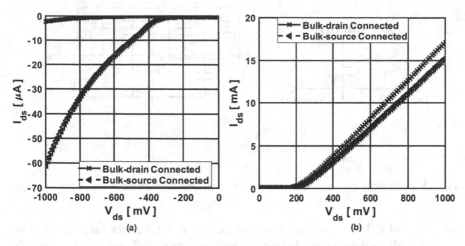

Fig. 5.2 Voltage–current relationship, (**a**) negative phase of input voltage swing, (**b**) in positive phase of input voltage swing

shorting the body-drain diode significantly improving the performance of the circuit. A comparison of the bulk-drain connection and the bulk-source connection is shown in Fig. 5.2. In the negative phase, as shown in Fig. 5.2a, the leakage current from the bulk-drain connected transistor is less than the bulk-source connected transistor. In the positive phase, as shown in Fig. 5.2b, the current from the bulk-drain connected transistor is stronger than the bulk-source connected transistor.

5.2.2 Inductor Peaking

Figure 5.3 shows the inductor-peaked diode-connected transistor and its equivalent circuit model in the on-state. From (4.36), R_{on} is decreasing with increasing V_{gs}. The simple short connected wire from gate to drain can be replaced by an inductor L. The inductor L is chosen such that together with the stray capacitances C_{gs} and C_{gd} (Fig. 5.3), an LC-resonator at the input RF frequency is formed. Therefore, the magnitude of voltage across the inductor and the capacitors is Q times higher than the input voltage. Therefore also the voltage swing V_{gs} at the gate of the transistor is larger than the input voltage. If V_{gs} increases, R_{on} will decrease and reduce the insertion loss which will lead to higher efficiency. In order to choose the value of L, three operational conditions have to be considered: accumulation, depletion, and inversion. The different modes of operation cause C_{gs} value variation. The variation of the value of the capacitor under the large signal swing and also turn on and off states will cause distortion of the resonator formed by L, C_{gs}, and C_{gd}. In most applications, this situation is not preferred. However, in the rectifier we can take

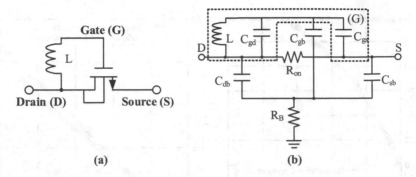

Fig. 5.3 (a) Inductor-peaked diode-connected transistor, (b) circuit model in the on-state

advantage of it. In the on-state, the resonator will make it easier for the current to pass through. In the off state, the current will have more difficulties to pass, which will improve the reverse isolation. The second order harmonics component can be described as $HD_2 = \frac{V_{in}}{4 \times V_t}$ [4]. The increase of V_{gs} also increases the second order harmonic, which in turn improves the rectification efficiency.

5.2.3   Output Filter

In the normal diode-connected transistor the gate and the drain are simply connected using a wire. If this method is applied at high frequency, a low-impedance path is formed by this wire and C_{gs} between the input and the output. In the rectifier application, the output is connected via a large capacitor to ground and can thus be treated as RF-ground. Therefore, next to the current path through R_{on}, there is a second path charging and discharging the storage capacitor. When the input voltage swing is higher than the output voltage, the current ($I_{charge1}$) is from input to the output. When the input voltage is lower than the output voltage, the current ($I_{charge2}$) is from output to the input. The current over one period is shown in Fig. 5.4, and can be expressed as

$$I_{avg} = \frac{\sum\limits_{t_1}^{t_2} I_{charge1} - \sum\limits_{t_2}^{t_3} I_{charge2}}{(t_2 - t_1) + (t_3 - t_2)} \tag{5.1}$$

$\sum\limits_{t_2}^{t_3} I_{charge2}$ is larger than $\sum\limits_{t_1}^{t_2} I_{charge1}$, because V_{out} is not lower than the 0. Thus, the direction of the I_{avg} is from the output to the input, and this current generates the discharging effect. Thus, this discharging will deteriorate the isolation and lead to low efficiency. In order to solve the problem of the high frequency signal

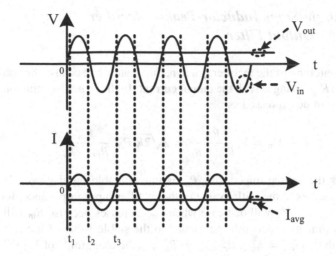

Fig. 5.4 Discharging current (I_{avg}) introduced by the signal (V_{in}) feed-through C_{gs}

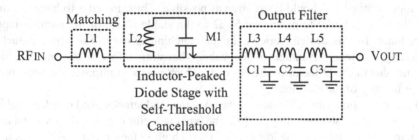

Fig. 5.5 Schematic of the 62 GHz one-stage inductor-peaked rectifier

path through the parasitic capacitor, a filter is necessary at the output, as shown in Fig. 5.5. It will filter out the high frequency component so as to prevent the discharging through the high frequency parasitic path.

5.3 mm-Wave Rectifier Implementation and Measurement

In this section, three types of mm-wave rectifiers are presented. The first one is the single-stage inductor-peaked rectifier with self-threshold cancelation and output filter. The second one is the three-stage inductor-peaked rectifier, with the inductor-peaked technology applied to the first and third stage. The third one is the three-stage rectifier with local threshold modulation applied to all three stages.

5.3.1   Single-Stage Inductor-Peaked Rectifier
with Output Filter

The input matching of the rectifier is a dynamic matching because the rectifier input impedance (R_{in}) changes with the output current. In the matched situation, the input voltage V_{in} can be expressed as

$$V_{in} = V_s \frac{R_{in}}{R_{in} + R_s} = 2\sqrt{2R_s P_{AV}} \frac{R_{in}}{R_{in} + R_s} \tag{5.2}$$

where R_s is the source impedance, P_{av} is the available input power. If the input matching matches to the initial value of the rectifier input impedance, for $i_{out} = 0$, $R_{in|inital} = R_{in| max}$, V_{in} will decrease when i_{out} increases because R_{in} will decrease. If the input matching network is matched to the stable value of the rectifier input resistance, that is $i_{out} = i_{max}$, $R_{in|stable} = R_{in| min}$, the magnitude of V_{in} will decrease because the input impedance of the rectifier will increase. In the latter case, it can take advantage of the mismatch between $R_{in|stable}$ and $R_{in|inital}$. For high sensitivity, the input voltage v_{in} should be as large as possible. Thus, in order to have a better sensitivity, it is better to match the 50 Ω to the stable phase of the rectifier input impedance. In this design, series inductor matching is used. The series inductor matching can take advantage of the voltage-boosting effect at the input of the rectifier due to the Q-factor of the network, which will also improve the sensitivity and efficiency of the rectifier.

Figure 5.5 shows the schematic of the one-stage inductor-peaked rectifier, and its die photograph is shown in Fig. 5.6 [5]. The input inductor is realized by a metal line inductor with metal strip shielding to achieve a high Q factor and make the design compact. To verify the layout performance, electromagnetic (EM) software (Agilent Momentum) has been used to simulate all the inductors, the interconnections, the bond pads, and the whole chip. Because all the inductors are close to each other, the return current through the ground metallization will degrade the performance. Therefore, the ground metals around the inductors are cut into small strips and each strip is perpendicular to the inductor line. The total chip area including bond pads is 0.52×0.35 mm^2.

The chip is measured by using on-wafer RF and DC probes. S-parameters are measured with an Agilent N5247A PNA-X network analyzer. Figure 5.7 shows the measured S_{11} of the inductor-peaked rectifier. At 62 GHz, the S_{11} is -32 dB. In the input power and output voltage (PV) characteristic measurement, the input power is provided by an Agilent E8267D and E8257D PSG signal generator. The input power is measured by a power sensor (Agilent E4491B) through a directional coupler (Agilent 83701E). The output voltage is measured through the Agilent 34401A digital multimeter. Figure 5.8 shows the measured input power output voltage (PV) characteristic at 61, 62, and 63 GHz. At 62 GHz, with -14.5 dBm input power, the rectifier provides 100 mV output voltage. In the rectifier efficiency measurement, in order to measure the output power accurately, an Agilent E5270B 8-slot precision

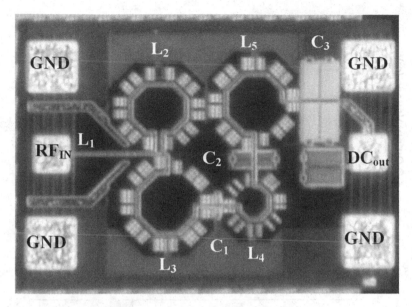

Fig. 5.6 Die micrograph of the 62 GHz inductor-peaked rectifier

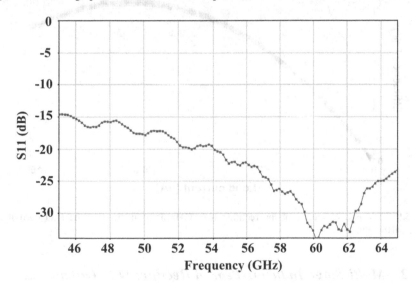

Fig. 5.7 Measured S_{11} of the proposed inductor-peaked rectifier

measurement mainframe is used as the load for the rectifier. The load current is provided and the corresponding output voltage is measured by the Agilent E5270B 8-slot precision measurement mainframe. Figure 5.9 shows the measured efficiency of the fabricated inductor-peaked rectifier. The maximum efficiency at 62 GHz is 7% for 1 mA current load.

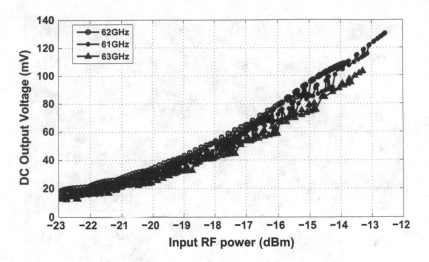

Fig. 5.8 Measured output voltage as a function of the input power

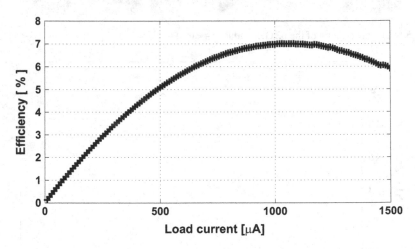

Fig. 5.9 Measured efficiency of the rectifier at 62 GHz as a function of the load current for −14.5 dBm input power

5.3.2   Multi-Stage Inductor-Peaked Rectifier with Output Filter

In the previous section, a one-stage inductor-peaked rectifier is presented with 7% efficiency. In mm-wave RFID applications [6] and mm-wave radio-triggered wake-up radios [7], a high output voltage is required to provide sufficient supply voltage for the active components, in the order of 1 V for 65 nm CMOS technology. The Dickson structure based voltage rectifier [8] is widely used to multiply the output voltage. The efficiency of the traditional Dickson structure based rectifier

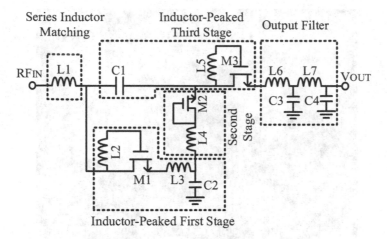

Fig. 5.10 Schematic of the three-stage inductor-peaked rectifier

at mm-wave frequency is around 1%, and the sensitivity is limited. By cascading the single-stage inductor-peaked rectifier, it can achieve the function of high performance multi-stage rectifier.

In Fig. 5.10, the three-stage inductor-peaked rectifier is presented. The first and third stage are formed by an inductor-peaked diode-connected transistor. To achieve a compact layout with short interconnects, the second stage is formed by a normal diode-connected transistor. In order to prevent the high frequency signal feed-through via the parasitic capacitor C_{gs}, which will discharge the load capacitor directly, there is a low pass filter in each stage of the rectifier. L_3, L_4 form the low pass filter in the first stage and second stage. The input matching of the rectifier is a dynamic matching because the rectifier's input impedance changes with the output current. For better sensitivity, the input voltage swing should be as large as possible. Thus, in order to have a better sensitivity, it is better to match the 50 Ω to the stable phase of the rectifier input impedance to take advantage of the mismatch. In Fig. 5.10, L_1 is the input series inductor matching stage. Besides impedance matching, the inductor L_1 also leads to a voltage-boosting effect at the input of the rectifier, which will also improve the sensitivity and efficiency of the rectifier.

Figure 5.11 shows the die photograph of the three-stage inductor-peaked rectifier. The chip is fabricated in 65-nm CMOS technology. The total chip area including bond pads is $0.48 \times 0.47 \, \text{mm}^2$. The input inductor, L_1, is realized by a metal line inductor with metal strip shield to achieve a high Q factor and make the design compact. Considering the coupling between L_3 and L_5, and the distance from C_2 to M_2, L_4 is also realized by an inter-stage metal line inductor with strip shielding, as shown in Fig. 5.11. Just as for the single-stage rectifier, also the three-stage rectifier performance was optimized by electromagnetic simulations. To verify the layout performance, electromagnetic (EM) software (Agilent Momentum) has been used to simulate all the inductors, the interconnections, the bond pads, and the whole chip.

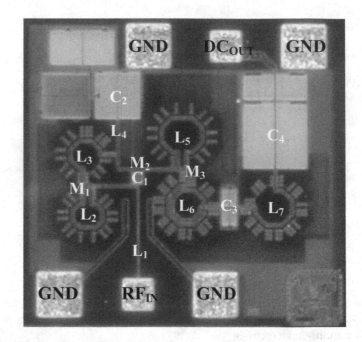

Fig. 5.11 Die micrograph of the three-stage inductor-peaked rectifier

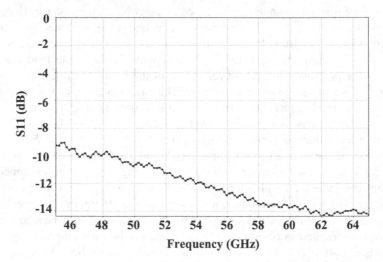

Fig. 5.12 Measured S_{11} of the three-stage inductor-peaked rectifier

The chip is measured with on-wafer RF and DC probes. S-parameters were measured with an Agilent N5247A PNA-X network analyzer. Figure 5.12 shows the measured S_{11} of the inductor-peaked rectifier from 45 to 65 GHz. Because of the equipment limitations, the s-parameters could not be measured at the

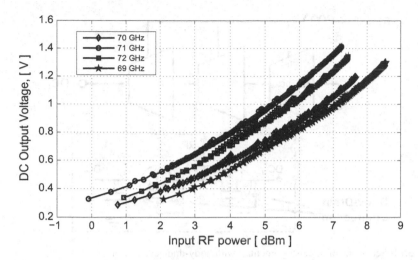

Fig. 5.13 Measured output voltage as a function of the input power

higher frequency range. In the input power and output voltage (PV) characteristic measurement, Fig. 5.13, the input power is provided by an Agilent E8267D and E8257D PSG signal generator. The input power is measured by a power sensor (Agilent E4491B) through a directional coupler (Agilent 83701E). The output voltage is measured through the Agilent 34401A digital multimeter. In the rectifier efficiency measurement, in order to measure the output power accurately, an Agilent E5270B 8-slot precision measurement mainframe is used as the load for the rectifier. The load current is provided and the corresponding output voltage is measured by the Agilent E5270B 8-slot precision measurement mainframe. The measured maximum efficiency at 71 GHz is 8.2% with 720 μA current load.

5.3.3 50 ∼ 60 GHz Broadband Rectifier

In the monolithic sensor node, a compact solution of the rectifier is required. Transistors with the local threshold voltage can achieve a smallest area solution. In this solution, no inductor is required to achieve a broadband performance. Figure 5.14 shows the schematic of the three-stage rectifier and its die photograph is shown in Fig. 5.15 [9]. The chip is fabricated in 65 nm CMOS technology. The total chip area including the matching inductor is $0.1 \times 0.146\,\text{mm}^2$. The series inductor matching is used to take advantage of the voltage-boosting effect at the input of the rectifier, which will further improve the sensitivity and efficiency of the rectifier. The patterned ground shielding is designed under the coupling capacitor, in order to decrease power loss to substrate resistance.

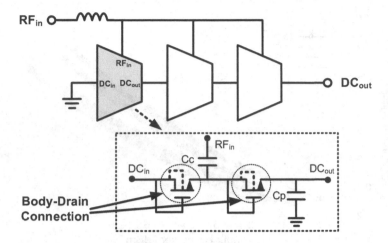

Fig. 5.14 Schematic of three-stage rectifier with body-drain connection

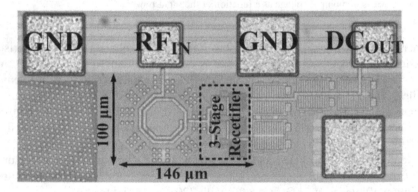

Fig. 5.15 Die micrograph of the three-stage rectifier with body-drain connection

The chip is measured with on-wafer RF and DC probes. S-parameters have been measured with an Agilent N5247A PNA-X network analyzer. Figure 5.16 shows the measured S_{11} of the three-stage rectifier from 30 to 67 GHz. In the input power and output voltage (P_{in}–V_{out}) characteristic measurement, shown in Fig. 5.17, the input power is provided by an Agilent E8267D and E8257D PSG signal generator. The input power is measured by a power sensor (Agilent E4491B) through a directional coupler (Agilent 83701E). The output voltage is measured through the Agilent 34401A digital multimeter. With 0 dBm input power, this rectifier can provide as high as 4 V output voltage at 52 GHz; at 60 GHz, the output voltage is 1.2 V. Over the $50 \sim 60$ GHz range, the circuit can provide 1 V output voltage with an input power lower than −2 dBm, as shown in Fig. 5.18. The rectifier achieves peak sensitivity at 52 GHz, requiring an input power of −7 dBm for an output voltage of 1 V. In the efficiency measurement, the load current is provided and the corresponding

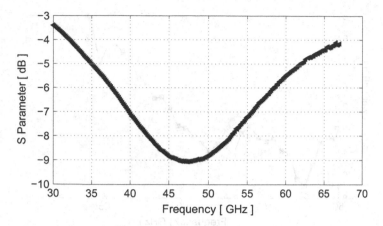

Fig. 5.16 Measured S_{11} of the rectifier

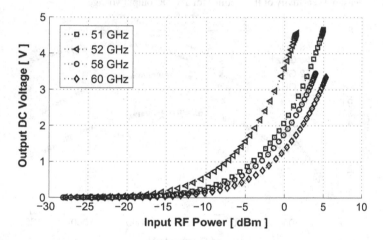

Fig. 5.17 Measured output DC voltage vs. input RF power characteristic of the rectifier

output voltage is measured by the Agilent E5270B 8-slot precision measurement mainframe. The peak efficiency of 8% is obtained at 50 GHz, with an 800 μA load current, as shown in Fig. 5.19.

5.4 Conclusions

In this chapter, the issues limiting the efficiency and sensitivity of the mm-wave rectifier are analyzed and several solutions to solve the problems are provided. Basically, there are three issues. The first one is related to transistor insertion loss, the second one is the isolation from input to the output, and the third is

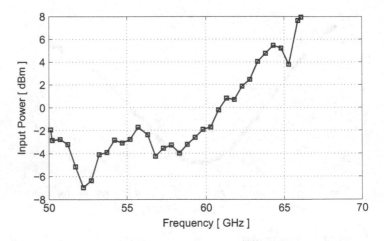

Fig. 5.18 Measured sensitivity of the rectifier for 1 V DC output voltage

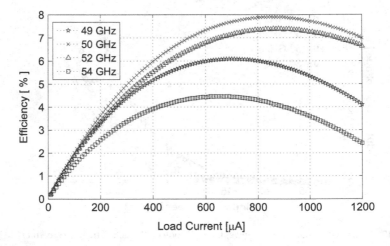

Fig. 5.19 Measured rectifier efficiency

the signal coupling to the lossy substrate through parasitic capacitors. In order to solve the problems coming from the transistor insertion loss, the methods of local threshold voltage modulation and inductor peaking are proposed. In order to solve the isolation problems, the transistor needs to be chosen small enough and an output filter is added. The loss introduced by the lossy substrate could be solved by optimizing the transistor layout so as to minimize the effect from the parasitic capacitor. Using the proposed methods, three rectifiers are implemented in 65 nm CMOS technology. The summary and comparison of the mm-wave rectifiers is presented in Table 5.1. By using the inductor-peaking method, local threshold modulation and an output filter, the single-stage inductor-peaked rectifier achieves 7% efficiency. In the first version of the three-stage rectifier design, the first stage

Table 5.1 Summary and comparison of mm-wave rectifiers

	JSSC 2010 [1]	A-SSCC 2012 [2]	Single-stage rectifier (Sect. 5.3.1)	Multi-stage rectifier (Sect. 5.3.2)	Broadband rectifier (Sect. 5.3.3)
CMOS technology (nm)	90	65	65	65	65
Area (mm^2)	1.235	0.06	0.182	1.09	0.0146
RF harvesting frequency (GHz)	45	60	62	70 to 72	50 to 60
Rectifier efficiency	1.2% @ 45 GHz	NA	7% @ 62 GHz	8% @ 71 GHz	8% @ 50 GHz
Sensitivity for 1 V output DC	2 dBm @ 45 GHz	2 dBm @ 60 GHz	NA	5 dBm @ 71 GHz	−7 dBm @ 52 GHz
					−2 dBm @ 60 GHz

and the third stage applied the inductor-peaked method. This three-stage rectifier reaches 8% efficiency at 71 GHz. With 5 dBm input power, the rectifier provides an output DC voltage of 1 V. In the second version of the three-stage rectifier design, the local threshold modulation technology is applied to all three stages, and this rectifier achieves −7 dBm input sensitivity for an output voltage of 1 V at 52 GHz. The third version of the broadband rectifier achieves the smallest size with only local threshold voltage modulation method. Compared with the traditional Dickson rectifier presented in [1, 2], the proposed method could increase the efficiency from 1.4% [1] to 8%, and the sensitivity is increased from 2 dBm [2] to −7 dBm for 1 V output voltage. In the future, the loss introduced from the lossy substrate could also be taken into account by adapting high resistivity substrate material.

References

1. S. Pellerano, J. Alvarado, Y. Palaskas, A mm-wave power-harvesting RFID tag in 90 nm CMOS. IEEE J. Solid-State Circuits **45**(8), 1627–1637 (2010)
2. S. Kawai, T. Mitomo, S. Saigusa, A 60 GHz CMOS rectifier with −27.5 dBm sensitivity for mm-Wave power detection, in *2012 IEEE Asian Solid State Circuits Conference (A-SSCC)*, November 2012, pp. 281–284
3. H. Gao, M. Matters-Kammerer, P. Harpe, P. Baltus, A 50–60 GHz mm-wave rectifier with bulk voltage bias in 65-nm CMOS. IEEE Microw. Wirel. Compon. Lett. **26**(8), 631–633 (2016)
4. P. Wambacq, W. Sansen, *Distortion Analysis of Analog Integrated Circuits* (Springer Science and Business Media, Boston, 2013)
5. H. Gao, M. Matters-Kammerer, D. Milosevic, A. van Roermund, P. Baltus, A 62 GHz inductor-peaked rectifier with 7% efficiency, in *2013 IEEE Radio Frequency Integrated Circuits Symposium (RFIC)*, June 2013, pp. 189–192
6. P. Pursula, T. Vaha-Heikkila, A. Muller, D. Neculoiu, G. Konstantinidis, A. Oja, J. Tuovinen, Millimeter-wave identification – a new short-range radio system for low-power high data-rate applications. IEEE Trans. Microw. Theory Tech. **56**(10), 2221–2228 (2008)

7. H. Gao, Y. Wu, M. Matters-Kammerer, J.-P. Linnartz, A. van Roermund, P. Baltus, System analysis and energy model for radio-triggered battery-less monolithic wireless sensor receiver, in *2013 IEEE International Symposium on Circuits and Systems (ISCAS)*, May 2013, pp. 1572–1575
8. J. Dickson, On-chip high-voltage generation in MNOS integrated circuits using an improved voltage multiplier technique. IEEE J. Solid-State Circuits **11**(3), 374–378 (1976)
9. H. Gao, M. Matters-Kammerer, D. Milosevic, A. van Roermund, P. Baltus, A 50∼60 GHz rectifier with −7 dBm sensitivity for 1 V DC output voltage and 8% efficiency in 65-nm CMOS, in *2014 IEEE MTT-S International Microwave Symposium (IMS)*, June 2014, pp. 1–3

Chapter 6
mm-Wave Monolithic Integrated Sensor Nodes

Abstract This chapter presents the analysis, implementation, and measurement of two fully integrated mm-wave temperature sensor nodes with on-chip antennas. These two sensor nodes provide two solutions for integrating a sensor node with on-chip antenna(s). The first solution contains two on-chip antennas, one for Tx and one for Rx, in which Tx/Rx have separate antennas. The second solution is a one on-chip antenna solution, in which the on-chip antenna is reused by Tx/Rx through an RF switch. These sensor nodes are implemented in 65 nm CMOS technology. The first sensor node contains a monopole antenna at 71 GHz for RF power harvesting, a storage capacitor array, an End-of-Burst monitor, a temperature sensor, and an ultra-low-power transmitter at 79 GHz. At 71 GHz, the RF to DC converter achieves a power conversion efficiency of 8% for 5 dBm input power. The second sensor node contains an integrated antenna, an RF switch, an on-chip wireless power receiver, and a temperature-correlated ultra-low-power transmitter. It measures only 1.83 mm^2 in 65 nm CMOS and weighs 1.6 mg. With the on-chip 30/65 GHz dual-frequency antenna and a three-stage inductor-peaked rectifier, the node can be wirelessly charged to 1.2 V. The output frequency of the temperature-correlated transmitter varies from 78.92 to 78.98 GHz, with a slope of 1.4 MHz/°C.

6.1 Introduction

Future wireless sensor networks require reliable, battery-less, miniaturized, low-cost sensor nodes with ultra-low-power consumption. Remote RF-powering provides a reliable, wireless source of power for such monolithically integrated sensor nodes [1]. RF-energy harvesting and data transfer at mm-wave frequencies will enable CMOS on-chip antenna integration. Combining highly integrated ultra-low-power mm-wave sensor nodes with wireless power transfer (WPT) with an on-chip antenna is a path towards battery-less, fully monolithically integrated, millimeter-sized sensor nodes with only a few milligram of weight [2]. Moreover, an important advantage of mm-wave wireless energy transfer is the higher received power level that can be achieved through a much narrower and highly directive beam [3] than would be created by a similar-sized phased array at lower frequencies. Therefore,

© Springer International Publishing AG 2018

H. Gao et al., *Batteryless mm-Wave Wireless Sensors*, Analog Circuits
and Signal Processing, https://doi.org/10.1007/978-3-319-72980-0_6

going to mm-wave operation is a crucial step towards successful wireless sensor networks [4]. At 60 GHz, the antenna size can be reduced to only 0.6 mm on silicon while it is at least 3 cm for off-chip $\lambda/4$ antennas at UHF frequencies [5]. Furthermore, a phased-array system with beam-forming algorithm applied at the base-station can make the mm-wave WPT more efficient. mm-Wave WPT with an on-chip antenna results in sensor systems with extremely compact size ($1.83\,\text{mm}^2$) and low weight (1.6 mg) which makes it possible to fit the sensor system into an extremely small area. Such monolithically integrated nodes require no additional packaging or maintenance. The maximum range of this type of nodes critically depends on the efficiency of the mm-wave energy harvesting with on-chip rectifiers, on capturing sufficient energy for activating the node and on minimizing the required Tx power. Recent publications [6, 7] demonstrate RFID tags either working at lower frequencies, from HF (13.56 MHz) to UHF (860 MHz to 2.45 GHz), or with limited efficiency at mm-wave frequency (1.2% at 45 GHz with 2 dBm input power [8]).

In the previous chapters, we presented results about the mm-wave rectifiers. In this chapter, we will discuss two mm-wave sensor nodes with two solutions of integrated on-chip wireless power receiver and on-chip antennas.

6.2   System Description

In this section, the system behavior of the sensor node will be introduced, including the system time domain behavior and temperature detection method. Based on the system time domain behavior requirement, two system architectures are proposed and implemented. The first system architecture is with separate Tx and Rx on-chip antennas, and the second system architecture is with a single Tx-Rx on-chip antenna.

6.2.1   System Behavior Description

The mm-wave sensor nodes could be implemented as fully passive sensor nodes without battery by using wireless power transfer technology. It is wirelessly powered by the energy transfer from the base-station. That wirelessly transferred energy is received and rectified by an on-chip wireless power receiver. The time domain system function is shown in Fig. 6.1. The sensor node is charged by the wireless transmitted power (R_X). The enable signal (Enable) activates the sensor nodes only when the sensor node is charged to higher than the minimum required supply voltage (1.2 V) and when the input wireless transferred power signal ends. The sensor node exits the stand-by mode and starts transmitting by quickly discharging the

Fig. 6.1 Time domain description of the system function

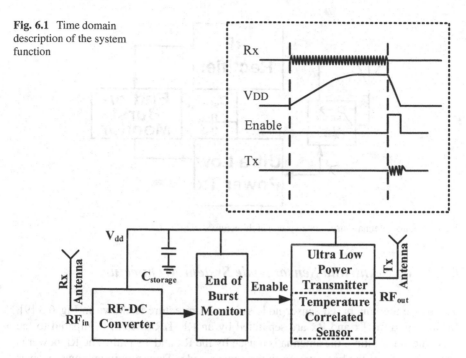

Fig. 6.2 Two-antenna sensor node system architecture

supply capacitor (V_{DD}). During discharging, the enable signal drops together with the supply voltage. The enable signal turns the system back to the stand-by mode when it is lower than the threshold voltage. The sensor node will wait for the next input power signal to recharge the sensor node and to transmit the next signal.

6.2.2 Two-Antenna Sensor Node System Architecture

The two-antenna based sensor node system architecture is shown in Fig. 6.2. In this solution, the Tx and Rx are separated by two separate antennas. One for the RF power receiver, and the other for the RF signal transmitter. The RF power is wirelessly transmitted from the base-station to the sensor node. First the Rx antenna receives the RF power and feeds it to the input of the RF energy receiver unit. The RF-DC converter rectifies the incoming RF power signal and the energy is stored on the on-chip storage capacitor. This energy is used later to power the Tx. The End-of-Burst monitor detects when the incoming power signal ends and turns the Tx part on. Then the Tx transmits a signal back through the (Tx) antenna.

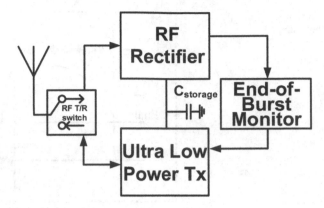

Fig. 6.3 One-antenna sensor node system architecture

6.2.3   One-Antenna Sensor Node System Architecture

The single antenna based sensor node system architecture is shown in Fig. 6.3 [9]. In this solution, Tx and Rx are separated by an RF T/R switch. Compared to the two-antenna solution, the antenna is reused by the Rx and Tx path. The RF power is transmitted from the base-station to the sensor node. The on-chip antenna receives the RF power, and feeds it to the T/R switch. In the initial phase, the T/R switch is connected to the RF-DC convertor. The generated energy from the RF power signal is stored in a storage unit. After the input power signal is over, the active part was turn on by the End-of-Burst Monitor and the T/R switch is connected to the Tx part. The active Tx transmit a temperature-correlated signal back through the on-chip antenna.

6.2.4   Comparison of the Two Solutions

The two-antenna solution is simple and straightforward. There is no other component inserted between the antenna and the wireless power receiver, therefore there will be no extra power loss, which could decrease the rectification efficiency. But the two on-chip antennas will occupy more area, and could easily introduce a coupling between them. In the one on-chip antenna solution, the system will be more complex, because the extra RF switch is required. This RF switch will generate extra loss which could decrease the system rectification efficiency. But the one-antenna solution will save chip area and there is no antenna coupling problem.

6.3 Circuit Design

In the wirelessly powered sensor nodes, there are several important components. Those are the on-chip antenna, the multi-stage rectifier, the end-of-burst monitor, and the ultra-low-power temperature-correlated transmitter. In the one on-chip antenna solution, the RF switch is also an important part. Also the matching between the antenna and the rectifier is an issue. In this section, each of the components will be analyzed.

6.3.1 Multi-Stage Rectifier for Wireless Power Receiver

In a wireless sensor node, the output voltage from the wireless power receiver should be high enough to power up the sensor node. A supply voltage of 1 V is a typical value for the active part to become functional in 65 nm CMOS technology. A multi-stage rectifier is used as the on-chip wireless power receiver to generate the supply output voltage. In Fig. 6.4 the three-stage inductor-peaked rectifier is presented. The first and third stages are formed by the inductor-peaked diode-connected transistor with local threshold voltage modulation. To achieve a compact layout with short interconnects, the second stage is formed by a normal diode-connected transistor. In order to prevent high frequency signal feed-through from the parasitic capacitor C_{gs}, which will discharge the load capacitor directly, there is a low pass filter in each stage of the rectifier. L_3, L_4 form the low pass filter in the first stage and second stage to prevent the RF signal feed-through to the DC output. The details of this multi-stage rectifier have been provided in Chap. 5.

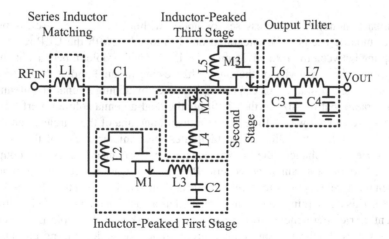

Fig. 6.4 Simplified schematic of three-stage inductor-peaked rectifier

Fig. 6.5 Simplified
schematic of End-of-Burst
monitor

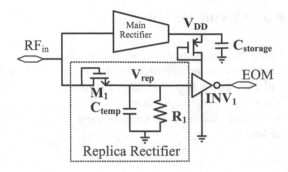

6.3.2   End-of-Burst Monitor

The end-of-burst monitor is embedded into the first stage of the multi-stage rectifier. The simplified schematic of the End-of-Burst Monitor is shown in Fig. 6.5. The End-of-Burst Monitor (EBM) is formed by a replica rectifier (M_1, C_{temp}), with a resistor load (R_1) and an inverter (INV_1). During the wireless charging phase, the replica rectifier generates a separate DC output voltage (V_{rep}) on the capacitor C_{temp}. As long as the input RF signal is on, the DC voltage generated through the inverter will stay low enough to keep the sensor node in the stand-by mode. After the RF power burst ends, V_{rep} is quickly discharged through the resistor load, R_1. This discharge is faster than the voltage drop of V_{DD}. This speed difference triggers the inverter which enables the ultra-low-power transmitter and sets the T/R switch to the Tx position.

6.3.3   RF Switch

In compact monolithic wireless sensor nodes without battery, a wireless power receiver module needs to be integrated on-chip together with the ULP radio. The decoupling between the Rx and Tx modules is a crucial problem, because it should not decrease the sensitivity of the energy harvesting module and at the same time keep the whole die small. This can be implemented by adding a separate antenna for the ULP radio, but the size of the die limits the dual-antenna solution performance [10]. A more elegant way is to reuse the transmit antenna of the wireless sensor for energy harvesting. The latter option facilitates the miniaturization of the wireless sensor node and enhances the on-chip antenna performance because no coupling between two on-chip antennas occurs. But, in this way, a decoupling function between the wireless power receiver module and the wireless data transfer module is required. This decoupling can be implemented as a dual-band antenna [11] with two different carrier frequencies so the matching network will decouple the modules, but this will make the system architecture complex. This decoupling can also be implemented by a diplexer[12], so the RF wireless receiver module and the wireless data transfer module can be located in different frequency bands and share the same

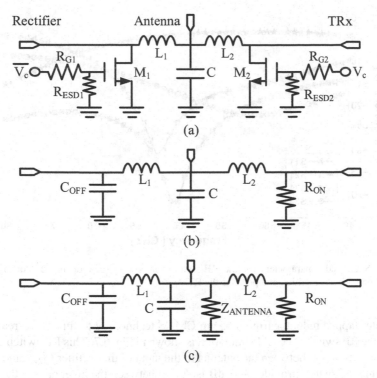

Fig. 6.6 Schematic of SPDT switch with its small-signal equivalent circuit for power harvesting mode. (**a**) Schematic of SPDT RF switch. (**b**) Small-signal equivalent circuit. (**c**) Small-signal equivalent circuit of isolation

antenna, but the diplexer area is large. Finally, the decoupling can also be achieved by using an RF switch to electrically separate the modules. Compared to the other solutions, the solution with RF switch is compact and requires a small area. It is also suitable for the low power application because of the simple system architecture.

The RF switch can be implemented using traveling-wave or on-chip $\lambda/4$ transmission lines but then it occupies a large die area. Additionally its insertion loss limits the on-chip wireless power receiver sensitivity. The proposed single-pole single-throw (SPST) RF switch is shown in Fig. 6.6a. It is constructed by two asymmetric SPST switches in back-to-back connection [13]. One side of the SPDT forms the path for wireless power reception while the other side is for wireless data communication. The resistors RESD are added at the gate of the switch transistors to enhance the breakdown characteristics of the transistor. In the wireless power receive mode, transistor M_1 is mostly equivalent to a capacitor, C_{off}, and transistor M_2 forms a resistance, R_{on}, because V_C is pulled up to high voltage while $(\overline{V_C})$ is pulled down to a low voltage. Thus, a matching network is achieved between the antenna and the input of the rectifier, and TRx is isolated from the antenna, as shown in Fig. 6.6b. The two matching inductors L_1 and L_2 can be jointly implemented as

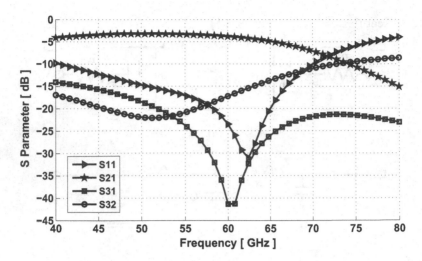

Fig. 6.7 Simulated S-parameters for the RF switch for power reception mode. Port 1 is the antenna, port 2 is the rectifier, and port 3 is the RTx

the center-tapped inductor from a 65 nm CMOS technology to further decrease the size of the RF switch. The EM simulation is shown in Fig. 6.7. This RF switch shows -2 dB insertion loss between the antenna to the input of the rectifier (S_{21}), and at the same time the switch provides -40 dB isolation between the antenna and the input of TRx (S_{31}). The isolation between rectifier and TRx port (S_{32}) is -18 dB.

6.3.4   On-Chip Antenna

At mm-wave frequency, the antenna can be implemented on-chip due to its small size, and this is a crucial step to achieve fully integrated sensor nodes. However, severe losses in the silicon substrate due to substrate modes propagating through the low-resistive silicon strongly limit the gain of on-chip antennas. The bulk substrate can be considered as a slab waveguide [5]. The cut-off frequencies of the transverse electric (TE) mode and transverse magnetic (TM) mode of a dielectric slab with infinite lateral extension placed in air can be determined from

$$f_c^{(n)} = \frac{nc_0}{2d\sqrt{\varepsilon_r - 1}} \tag{6.1}$$

where n is the mode number, $n \in N_0$, c_0 is the speed of light in vacuum, and d and ε_r are the thickness and permittivity of the slab's material, respectively [14]. Thus, the 0th order mode has a zero Hz cut-off frequency and therefore always gets excited. Furthermore, since silicon chips are of finite size, the excited substrate waves get partly reflected and partly radiated at the chip edges. The radiated part superimposes on the waves that are directly radiated from the antenna and, hence strongly affects the antenna radiation pattern. Therefore, the exact overall pattern depends to a large extent on the dimensions of the chip. A simple measure that can be taken to enhance the off-chip radiation is the implementation of a metal plate covering the chip [5]. By this simple measure, the substrate can be considered as a grounded dielectric slab, also known as surface waveguide, with corresponding surface-waveguide modes. The lowest-order TM mode (TM_0-mode) still has 0 Hz cut-off frequency, i.e., it can still be excited. The lowest-order TE-mode (TE_1-mode), however, exhibits a non-zero cut-off frequency. Hence, for frequencies below that cut-off, all guided TE-modes are suppressed and the radiation pattern and efficiency are improved [9]. For a 60 GHz antenna design, for example, the thickness of the silicon substrate needs to be below 350 μm in order to avoid the propagation of TE-modes up to 66 GHz. In this design, top metal was chosen as the metal plate, as shown in Fig. 6.8. In order to reduce the amount of lossy silicon around the monopole and the coupling from the sealing ring of the chip, it was placed at a distance of 50 μm from the edge of the silicon substrate, which has a thickness of 280 μm and an electric resistivity of 10 Ω/cm. In simulation, this on-chip monopole antenna shows −1.68 dBi antenna gain, as shown in the 60 GHz far-field simulation result in Fig. 6.8.

6.3.5 Matching Between the Rectifier and the On-Chip Antenna

An antenna is used to convert the energy from the field to electrical energy, and the rectifier is used as the on-chip wireless power receiver. The matching between those two components is important to achieve maximum rectification efficiency. In the sensor node, the antenna connects directly to the rectifier, which contains diodes that are sensitive to the voltage swing at the input terminals. Its characteristics affect the overall performance of the rectenna [15].

As shown in Fig. 6.9a, an antenna in general has a complex source impedance, which can be written in the form of $Z_{ant} = R_s + jX_{ant}$. The voltage amplitude of the source V_s is

$$V_s = 2\sqrt{2R_sP_{AV}} \qquad (6.2)$$

where P_{AV} is the available power from the antenna. In the matched situation, the antenna and the rectifier interface can be modeled as shown in Fig. 6.9b, and we assume the rectifier has a linear real input impedance, R_i. The input voltage V_{in} can be expressed as

$$V_{in} = V_s \frac{R_i}{R_i + R_s} = 2\sqrt{2R_sP_{AV}} \frac{R_i}{R_i + R_s} \qquad (6.3)$$

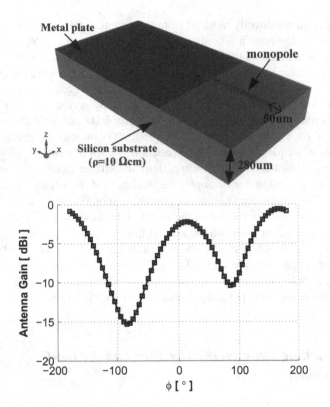

Fig. 6.8 3D EM model of 60 GHz on-chip antenna and its far-field (E-field) simulation result at 60 GHz

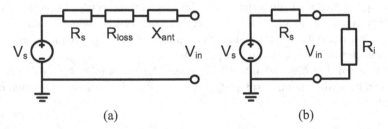

(a)   (b)

Fig. 6.9 Antenna rectifier interface. (a) Antenna equivalent model. (b) Antenna rectifier interface

From (6.2), V_s is proportional to the square-root of the radiation resistance R_s. Maximum transferred power is obtained when the rectifier is matched to the antenna. So, both the power matching and a high radiation resistance should be taken into consideration for the design.

In reality, the power matching is a dynamic matching because the rectifier input resistance is changed with the output current [16]. Therefore the first order linear model is not complete. This dynamic matching has been discussed in paragraph 5.3.1.

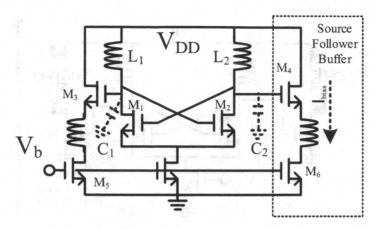

Fig. 6.10 LC oscillator with source follower buffer

6.3.6 Transmitter with Temperature Sensing

In the ultra-low power sensor node, an LC oscillator is used to generate the Tx signal at f_{Tx}, and its output buffer is used to drive the antenna directly. This method is simple enough to keep the whole system power consumption low. There are two types of output buffers, that could be suitable to drive the 50 Ω antenna input, as shown in Figs. 6.10 and 6.11. The first is the source follower, and the second is the common source buffer. The oscillator frequency is determined by the tank inductance (L_1, L_2) and capacitance (C_1, C_2). The tank capacitance is dominated by the device gate capacitance (M_1, M_2, M_3, M_4), which is tightly controlled in modern CMOS processes, but is sensitive to temperature [17]. Due to the different buffer structures, the parasitic capacitors of the transistor will cause different temperature dependencies. In the source follower structure, the bias current will become larger with temperature, which will increase the gate capacitor of M_3 and M_4. Because of the increase of the gate capacitor from the source follower, the frequency will drift to lower frequency with increasing temperatures. In the common source structure, with increasing temperature, the gate capacitor of M_1, M_2, M_3, and M_4 will become smaller. The decreasing tank capacitor value will increase the output frequency. Those two effects will be shown in the measurements of two types of sensor nodes.

6.4 mm-Wave Sensor Nodes Implementation

In this section, two versions of the sensor nodes are implemented in 65 nm CMOS technology. The first one is the sensor node with separate Tx and Rx antennas. The second one is the sensor node with a single Tx/Rx antenna.

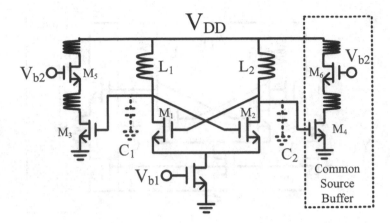

Fig. 6.11 LC oscillator with common source buffer

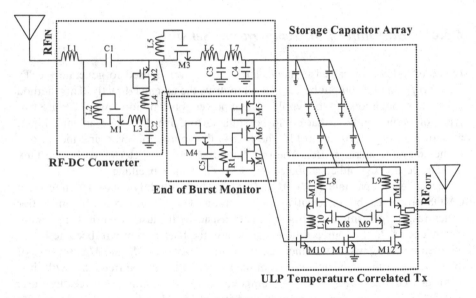

Fig. 6.12 Simplified schematic of the sensor node with two antennas, including the RF-DC converter, the End-of-Burst monitor, the storage capacitor, the ultra-low-power (ULP) temperature-correlated transmission stage (TX) and two on-chip antenna

6.4.1   mm-Wave Sensor Node with Two Antennas

Figure 6.12 shows the simplified schematic of the fully wireless node for the two-antenna solution. There are two on-chip antennas, one for the power receiver and the other for the transmitter. The DC voltage is generated by the three-stage inductor-peaked rectifier, and is stored on the on-chip capacitor array. In order to increase the

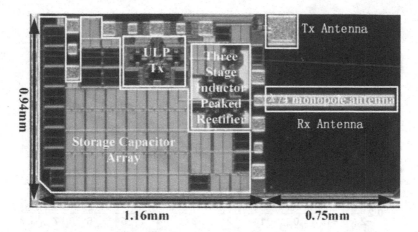

0.94mm

1.16mm   0.75mm

Fig. 6.13 Die micrograph of the sensor node with two on-chip antennas

capacitor density, a MoM capacitor was integrated below the MiM capacitor. The capacitor array is connected such that it has a low series resistance. The End-of-Burst monitor is formed by an inverter (M_5, M_6, M_7) and a small rectifier (M_4, C_5) with a small resistor load (R_1). The input of the monitor is attached to the input of the third stage of the rectifier. After the RF power burst ends, the End-of-Burst monitor generates the enable signal to the ultra-low-power transmit stage (Tx), consisting of a VCO and an output buffer. The 79 GHz oscillator generates the output burst when the enable signal is high. The complete sensor node was implemented in 65 nm CMOS technology and occupies 1.91×0.94 mm^2 of chip area, as shown in Fig. 6.13. Due to a manufacturing error, the loop antenna of the Tx part is fully filled with the metal inside, which makes it looks like a patch antenna. A wireless temperature measurement was unfortunately not possible for this sensor node.

The 71 GHz energy harvesting tag was implemented in a 65 nm CMOS technology and is shown in Fig. 6.13. The chip is measured with on-wafer RF and DC probes. Figure 6.14 shows the measured sensitivity of the three-stage inductor-peaked rectifier. With 5 dBm input power at 71 GHz to activate the tag, the output voltage is 1 V with 1 MΩ load. Figure 6.15 shows the measured efficiency of the three-stage inductor-peaked rectifier. The maximum efficiency at 71 GHz is 8% for a 700 μA current load. The minimum required supply voltage for the oscillator to start oscillation is 670 mV and the Tx current consumption is 4 mA. The Tx frequency is 79.1 GHz as shown in Fig. 6.16.

Figure 6.17 shows the charging and discharging of the tag. With 5 dBm input power at 71 GHz, the system is charged to 937 mV in 4.1 ms. When the RF input power signal ends, the tag fully discharges in 7 μs, as shown in Fig. 6.17. During discharging, the oscillator frequency is related to ambient temperature. From 32 °C to 80 °C, it shows a reasonably linear behavior between 79.12 and 78.88 GHz with slope $k = -22$ MHz/°C (Fig. 6.18).

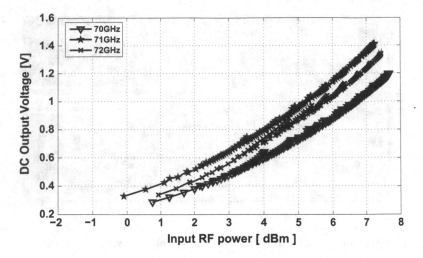

Fig. 6.14 Measured rectifier output voltage as a function of the RF input power (dual antenna node)

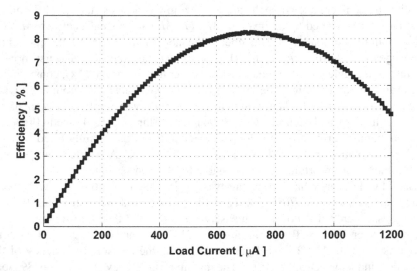

Fig. 6.15 Measured rectifier efficiency with load current at 71 GHz with 5 dBm input power (dual antenna node)

From the measured data of the charging time, 13 μJ energy is received from the antenna during 4.1 ms. The energy stored on chip is 0.44 nJ. The system converting efficiency is 0.003%, which is different from the measured rectifier efficiency. This is due to the loading effect and the leakage current during the charging time. During the charing time, the energy is lost through the leakage current, which makes the stored energy less than expected. From the measurement, it is also noticed that

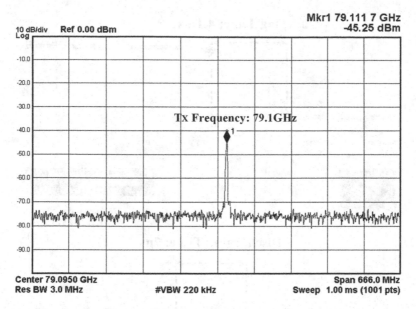

10 dB/div   Ref 0.00 dBm
Log

Mkr1 79.111 7 GHz
-45.25 dBm

Tx Frequency: 79.1GHz

Center 79.0950 GHz
Res BW 3.0 MHz   #VBW 220 kHz

Span 666.0 MHz
Sweep 1.00 ms (1001 pts)

Fig. 6.16 Measured Tx frequency of the ultra-low-power transmitter (dual antenna node)

the efficiency of the rectifier should be optimized to the averaged output current during charging, which is around 200 μA in this sensor node, for improved system converting efficiency.

6.4.2 mm-Wave Sensor Node with One Antenna

Figure 6.19 shows a die photograph of a single on-chip antenna sensor node with RF switch for dual-frequency operation. To achieve dual-frequency operation, the on-chip antenna is designed to be a $\lambda/4$ monopole antenna around 65 GHz and a $\lambda/8$ monopole antenna around 30 GHz. The T/R switch is controlled by the End-of-Burst detector. During the energy-harvesting phase, the T/R switch is connected to the input of a three-stage inductor-peaked rectifier that performs RF to DC conversion. The complete sensor node was implemented in 65 nm CMOS technology and occupies $1.85 \times 0.99\,\text{mm}^2$ of chip area [9].

Figure 6.20 demonstrates the wireless measurement of the system with on-chip RF energy receiver and wireless temperature sensing. The input RF power is fed to the sensor node through an open wave-guide antenna, the output DC voltage is measured by the voltage meter. The minimum required supply voltage for the oscillator is 750 mV and the Tx current consumption is 6 mA. At 30 GHz, with 16 dBm input RF power wirelessly fed to the on-chip antenna, the system successfully charged to 1.2 V. At 65 GHz, with 14 dBm input RF power wirelessly fed to the on-chip antenna, the system successfully charged to 1 V. During discharging, the oscillator

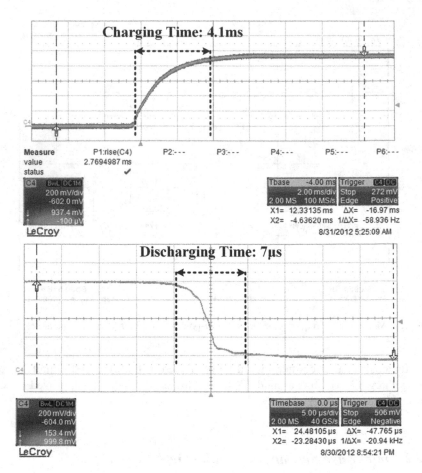

Fig. 6.17 Measured time domain tag charging and discharging with 5 dBm input RF power (dual antenna node)

frequency is related to the ambient temperature. The transmitted frequency was wirelessly detected, as shown in Fig. 6.21, and varies in a reasonably linear way between 78.92 and 78.98 GHz for a temperature range between 20 and 60°C, which corresponds to a slope of $k = 1.4\,\text{MHz}/°\text{C}$.

6.5 Conclusion

In this chapter, two architectures of the monolithic wireless sensor nodes are provided and two sensor node are implemented in the 65 nm CMOS technology. The first sensor node provides a solution with two on-chip antennas, one for receive and one for transmit mode. The second sensor node provides a solution for a

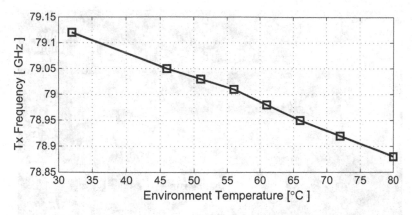

Fig. 6.18 Measured Tx frequency relationship to environment temperature (dual antenna node)

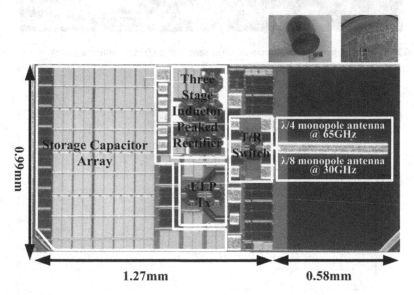

Fig. 6.19 Die micrograph of the sensor node with one on-chip antenna and RF switch

monolithic sensor node with one on-chip antenna, and it contains an RF switch and a dual-frequency antenna to achieve a one-antenna solution. Comparing the two versions, the second version can solve the coupling between two antennas to enhance the antenna performance. Table 6.1 summarizes the measured performance of the sensor nodes. The first sensor node provides best efficiency (8%) at 5 dBm input power. The second sensor node is a fully monolithically integrated sensor node including an on-chip antenna that can be wirelessly powered and sends the sensor information at mm-wave frequencies. Moreover, it has a 65 and 30 GHz dual-frequency energy-harvesting function which can make the system more flexible. The measurement indicates that the system conversion efficiency is different from the

Fig. 6.20 Wireless charging measurement of the sensor node (single antenna node)

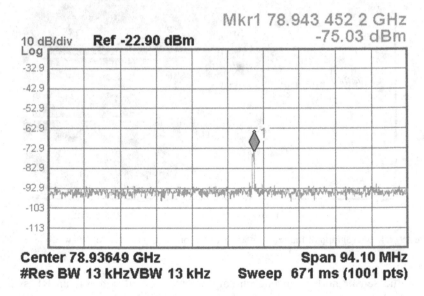

Fig. 6.21 Measured Tx frequency (single antenna node)

rectifier efficiency. It suggests that in the future, the leakage current during charging should be taken into consideration in order to optimize the rectifier efficiency according to that load current.

Table 6.1 Performance summary and comparisons

	ISSCC 2012 [6]	JSSC 2010 [8]	1st sensor node [10]	2nd sensor node	
Technology (nm)	65	90	65	65	
Area (mm²)	0.6	1.235	1.09	1.25	
Area with antenna	4.6 mm² (PCB etc. not incl.)	Not available	1.79 mm²	1.83 mm²	
Antenna on chip	No	No	Yes	Yes	
Sensor type	None	None	Temperature	Temperature	
RF energy frequency (GHz)	1.86	45	70~72	65	30
System efficiency	55%	1.2%	Not available	1.8%	0.7%
On-wafer measured power for waking up the node	−10 dBm 0.7 V	2 dBm 1 V	5 dBm 1 V	11 dBm 1 V	12.5 dBm 1 V
Power to wireless waking up the node	−10 dBm 0.7 V	Not available	Not available	13 dBm 1 V	15 dBm 1 V
Tx frequency	Not available	60 GHz	79 GHz	79 GHz	

References

1. F. Kocer, M. Flynn, A new transponder architecture with on-chip ADC for long-range telemetry applications. IEEE J. Solid-State Circuits **41**(5), 1142–1148 (2006)
2. H. Gao, Y. Wu, M. Matters-Kammerer, J.-P. Linnartz, A. van Roermund, P. Baltus, System analysis and energy model for radio-triggered battery-less monolithic wireless sensor receiver, in *2013 IEEE International Symposium on Circuits and Systems (ISCAS)*, May 2013, pp. 1572–1575
3. Y. Yu, P. Baltus, A. van Roermund, A. de Graauw, E. van der Heijden, M. Collados, C. Vaucher, A 60 GHz digitally controlled RF-beamforming receiver front-end in 65 nm CMOS, in *IEEE Radio Frequency Integrated Circuits Symposium, 2009. RFIC 2009*, June 2009, pp. 211–214
4. Y. Wu, J. Linnartz, H. Gao, P. Baltus, J. Bergmans, System study of a 60 GHz wireless-powered monolithic sensor system, in *2011 8th International Conference on Information, Communications and Signal Processing (ICICS)*, December 2011, pp. 1–5
5. U. Johannsen, A. Smolders, R. Mahmoudi, J. Akkermans, Substrate loss reduction in antenna-on-chip design, in *IEEE Antennas and Propagation Society International Symposium, 2009. APSURSI '09*, June 2009, pp. 1–4
6. G. Papotto, F. Carrara, A. Finocchiaro, G. Palmisano, A 90 nm CMOS 5 Mb/s crystal-less RF transceiver for RF-powered WSN nodes, in *2012 IEEE International Solid-State Circuits Conference Digest of Technical Papers (ISSCC)*, February 2012, pp. 452–454

7. H. Reinisch, M. Wiessflecker, S. Gruber, H. Unterassinger, G. Hofer, M. Klamminger, W. Pribyl, G. Holweg, A 7.9 μw remotely powered addressed sensor node using EPC HF and UHF RFID technology with −10.3 dBm sensitivity, in *2011 IEEE International Solid-State Circuits Conference Digest of Technical Papers (ISSCC)*, February 2011, pp. 454–456
8. S. Pellerano, J. Alvarado, Y. Palaskas, A mm-wave power-harvesting RFID tag in 90 nm CMOS. IEEE J. Solid-State Circuits **45**(8), 1627–1637 (2010)
9. H. Gao, M.K. Matters-Kammerer, P. Harpe, D. Milosevic, A. van Roermund, J.P. Linnartz, P.G.M. Baltus, A 60-GHz energy harvesting module with on-chip antenna and switch for co-integration with ULP radios in 65-nm CMOS with fully wireless mm-wave power transfer measurement, in *2014 IEEE International Symposium on Circuits and Systems (ISCAS)*, June 2014, pp. 1640–1643
10. H. Gao, M. Matters-Kammerer, P. Harpe, D. Milosevic, U. Johannsen, A. van Roermund, P. Baltus, A 71 GHz RF energy harvesting tag with 8% efficiency for wireless temperature sensors in 65 nm CMOS, in *2013 IEEE Radio Frequency Integrated Circuits Symposium (RFIC)*, June 2013, pp. 403–406
11. J.-H. Huang, J.-W. Wu, Y.-L. Chiou, C. Jou, A 24/60 GHz dual-band millimeter-wave on-chip monopole antenna fabricated with a 0.13-μm CMOS technology, in *IEEE International Workshop on Antenna Technology, 2009. iWAT 2009*, March 2009, pp. 1–4
12. B. Pan, Y. Li, G. Ponchak, M. Tentzeris, J. Papapolymerou, A low-loss substrate-independent approach for 60-GHz transceiver front-end integration using micromachining technologies. IEEE Trans. Microw. Theory Tech. **56**(12), 2779–2788 (2008)
13. J. He, Y.-Z. Xiong, Y.P. Zhang, Analysis and design of 60-GHz SPDT switch in 130-nm CMOS. IEEE Trans. Microw. Theory Tech. **60**(10), 3113–3119 (2012)
14. R.F. Harrington, *Time-Harmonic Electromagnetic Fields* (McGraw-Hill, New York, 1990)
15. H. Gao, U. Johannsen, M.K. Matters-Kammerer, D. Milosevic, A.B. Smolders, A. van Roermund, P. Baltus, A 60-GHz rectenna for monolithic wireless sensor tags, in *2013 IEEE International Symposium on Circuits and Systems (ISCAS2013)*, May 2013, pp. 2796–2799
16. J.-P. Curty, M. Declercq, C. Dehollain, N. Joehl, *Design and Optimization of Passive UHF RFID System* (Springer, New York, 2007)
17. D.A. Neamen, *Semiconductor Physics and Devices: Basic Principles* (McGraw-Hill, Boston, 2003)

Chapter 7
mm-Wave Low-Power Receiver

Abstract In the previous chapters, the rectifier for an on-chip wireless power receiver is analyzed and the mm-wave rectifier is implemented together with on-chip antenna(s) and temperature-correlated sensor in 65 nm CMOS technology to demonstrate the possibility to realize fully monolithic sensor nodes on silicon. In order to provide sensor nodes with more functionality, an ultra-low-power receiver must be co-integrated with the wireless power receiver module to receive commands from the base-station. In this chapter, a mm-wave ultra-low-power receiver architecture is proposed and studied. An injection-locked oscillator based architecture is proposed and implemented in 65 nm CMOS technology.

7.1 Introduction

Most wireless sensors operate on batteries, which limits their life-time. To overcome this limitation, a concept of a wireless sensor system with fully monolithic sensor nodes based on RF wireless power transfer is demonstrated in [1]: the sensors have antennas, transceivers, sensing and energy scavenging functions fully integrated into CMOS, which keeps their cost very low and makes them suitable for mass production and allows widespread deployment. However, because the energy that can be wirelessly received and stored on chip is rather limited due to the small size of the monolithic sensors and the limited efficiency of the rectifiers, all building blocks have to be extremely energy efficient. In wireless communication, an effective way to reduce the average power consumption is to make the transmitter operate discontinuously by interleaving active periods with long idle phases. Normally, two methods are widely used, namely periodic duty cycling [2] and adding an extra wake-up receiver (WuRx) to monitor the communication in low-power sleeping mode [3] and wake up the receiver to active high-power mode only when necessary. Both methods are effective in reducing power (energy) consumption of wireless sensors, however both require power during the idle time, which is not possible for the targeted fully passive sensor nodes.

© Springer International Publishing AG 2018   79
H. Gao et al., *Batteryless mm-Wave Wireless Sensors*, Analog Circuits
and Signal Processing, https://doi.org/10.1007/978-3-319-72980-0_7

In this chapter, an architecture for radio-triggered passive wireless sensor nodes is introduced. The sensor stays passive and does not consume any power during idle time. It is activated on-demand by a base-station, which transmits energy wirelessly to trigger and charge up the sensor node. Therefore the sensor only consumes energy when it is being polled. Compared with a duty-cycled solution, it will not miss a call from the base-station. Compared with a WuRx solution, there is no always-on radio, so it has better energy efficiency. In this chapter, the receiver operation of such a sensor node is studied and its receiver is named the radio-triggered passive receiver (RTRx). The energy models that relate power (energy) consumption and key performance figures (e.g., efficiency, noise figure, gain, etc.) are presented, for all the essential building blocks of the receiver, including the antenna, the matching network, the RF rectifier, the LNA, the oscillator, and the mixer. The study shows that the 60 GHz mm-wave band is suitable for the targeted specifications. Moreover, a system evaluation of the sensor receiver in 65 nm CMOS technology is presented to demonstrate the achievable performance with the proposed receiver architecture. In the end, a design and measurement results of a 60 GHz ultra-low-power receiver is presented in 65 nm CMOS technology.

7.2   Radio-Triggered Passive Receiver Architecture

In this section, the system architecture of the proposed radio-triggered passive receiver is presented. As shown in Fig. 7.1, the receiver consists of two main modules, namely the energy scavenging module used to rectify the received RF energy from the base-station and the receiving module, which is used to receive information from the base-station. In the energy receiver module, the received RF signal from the integrated antenna passes first through a matching network to ensure maximum power transfer to the subsequent parts of the circuits. The RF signal is then converted to a DC signal by an RF rectifier and the energy is stored in the energy storage unit, which could be an on-chip capacitor. Once enough energy is

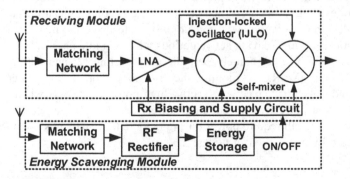

Fig. 7.1 The proposed radio-triggered passive receiver architecture

stored and the period of wireless power transfer is finished, the receiving module is triggered on to start receiving. In the receiving module, a similar matching network is implemented and a low noise amplifier (LNA) is used to amplify the incoming signal. On-Off-Keying (OOK) is used in low-power sensor applications for its simple demodulation circuit [4]. An OOK demodulator is implemented by a self-mixer, consisting of an injection-locked oscillator (IJLO) which locks its oscillating frequency to the incoming RF signal and a passive mixer.

Compared with other ultra-low-power receiver architectures [4, 5], the proposed radio-triggered passive receiver requires no battery as the energy is provided by the on-chip wireless power receiver. Moreover, the receiving mode is only triggered on when sufficient energy is scavenged via the RF power transfer from the base-station. Therefore, it is a passive structure and does not consume any energy in the off mode.

7.3 Energy Models

The radio-triggered wireless sensor receiver is composed of an RX antenna, a matching network, an RF rectifier, an LNA, an IJLO, and a mixer. In this section, the energy models for these blocks in the analog chain of the receiver are presented. These energy models are then used to identify a suitable frequency band for the proposed system and to understand the design trade-offs among different blocks.

7.3.1 Antenna and Matching Network

In monolithic sensor nodes, all components, including the antenna, are integrated on chip. For two reasons the mm-wave frequency band is suitable for on-chip antenna integration. First, at millimeter wave frequencies, it is possible to integrate reasonably sized on-chip antennas within the limited sensor chip area. For example, in silicon at 60 GHz, a $\lambda/2$ dipole antenna has a length of 1.1 mm. The second reason is related to the efficiency of the matching network. The self-impedance of an antenna depends on the relative size of the antenna to the wavelength. For the same physical antenna size, the antenna impedance changes with the center frequency. With decreasing frequency, the relative size of the antenna decreases and the antenna impedance becomes capacitive. To match the antenna impedance with a typical input impedance of 50 Ω to the rest of the circuit, we assume a simple and robust L matching network shown in Fig. 7.2.

The antenna self-resistance and reactance are denoted by R_s and X_s and the load resistance is denoted by R_o. The capacitive part of the input impedance of the rectifier will be tuned out by the matching network. The non-ideality in the matching network is modeled with series resistances R_l and R_C added to the inductor and capacitor with reactance X_l and X_C, respectively. Assuming the loss is dominated by the loss from R_l, the efficiency of the matching network, defined by the ratio of

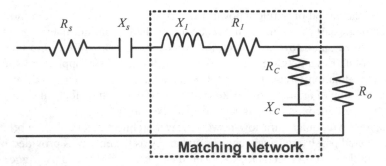

Fig. 7.2 A typical simple L matching network

the power available at the load for a lossy and an ideal matching network, can be expressed as

$$\eta_m\left(f\right) = \frac{P_L^{\text{lossy}}\left(f\right)}{P_L^{\text{ideal}}} = \left(\frac{2R_s\left(f\right)}{2R_s\left(f\right) + R_l\left(f\right)}\right)^2 = \left(\frac{2}{2 + \frac{Q(f)\pm Q_a(f)}{Q_l}}\right)^2 \qquad (7.1)$$

where $Q\left(f\right) = \sqrt{\frac{R_0}{R_s(f)} - 1}$, and $Q_a\left(f\right) = \left|\frac{X_s(f)}{R_s(f)}\right|$ are the Q factors of the matching network and antenna, respectively, and Q_l is the Q factor of the non-ideal inductor. Note that the "$\pm$" sign is "+" when the antenna is capacitive and is "−" when the antenna is inductive. From (7.1), it can be seen that with the small on-chip antenna, the efficiency of the matching network increases with frequency. The Q factor of a $\lambda/2$ line increases with the $\sqrt{f}$, because of the size reduction with $1/f$ and the skin effect with $\sqrt{f}$.

7.3.2 RF Rectifier

In the radio-triggered passive wireless receiver, the rectifier is the component that converts the incoming RF signal to DC for energy storage and triggering of the receiver operation. A Dickson structured multi-stage rectifier is used to provide sufficient output supply voltage. The rectifier power conversion efficiency (η_{rec}) is defined as the ratio of the output DC power ($\frac{1}{2}CV^2$) to the input RF power. A detailed analysis of the rectifier is presented in Chap. 4. In this section, curve fitting is adopted to build a numerical model for the rectifier for system level optimization. The numerical model is based on the data from simulated rectifier efficiency performance, which is presented in Chap. 5.3. The numerical model that approximates the efficiency vs. input power curve with a second order polynomial is expressed in (7.2).

$$\eta_{\text{rec}} \approx AP_{\text{in}}^2 + BP_{\text{in}} + C \qquad (7.2)$$

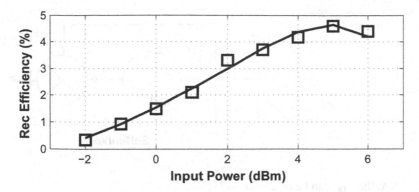

Fig. 7.3 Rectifier efficiency with input power

The efficiency of the Dickson structure rectifier for different input power levels using circuit simulations is plotted in Fig. 7.3. In the 60 GHz frequency band, the efficiency of the rectifier is described in (7.3).

$$\eta_{rec} (\%) \approx -0.66P_{in}^2 + 4.21P_{in} - 1.99 \tag{7.3}$$

Here P_{in} is the input power in mW. To obtain a system-level view of the frequency dependence of the energy scavenging module, we assume a base-station with transmit power P_t and physical antenna effective area A_e. The power at the output of the rectifier at a distance d can be expressed as

$$P_0(f) = \frac{P_t \left(4\pi A_e / \lambda^2\right) G_r \eta_m(f) \eta_{rec}(f)}{(4\pi d / \lambda)^2} = \frac{P_t A_e G_r \eta_m(f) \eta_{rec}(f)}{4\pi d^2} \tag{7.4}$$

where G_r is the receiver gain, and η_m and η_{rec} are the efficiency of the matching network and the rectifier, respectively. Here the on-chip antenna gain is assumed to be approximately 0 dBi for very small antennas, and hence the receive antenna gain G_r is approximately independent of frequency. We can see that the power at the rectifier output is frequency dependent mainly via $\eta_m(f)$ and $\eta_{rec}(f)$. Although using lower frequencies results in higher efficiency in rectifiers, the loss in the matching circuit and the antenna is much higher than the increase in rectifier efficiency. As a result, considering the size limitation of the monolithic sensor nodes, the 60 GHz band is a suitable frequency band for this application.

7.3.3 LNA

The specifications of the LNA include gain (G), noise factor (NF), third-order input intercept point (IIP_3), S-parameters, and DC power consumption (P_{DC}). For a single-

Fig. 7.4 Simplified
illustration of self-mixer

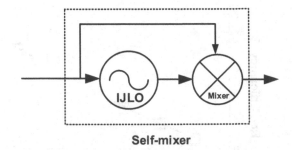

Self-mixer

stage LNA, the P_{DC} can be described as [6]:

$$P_{DC} = k \times G \times IIP_3 \times NF \tag{7.5}$$

where k is a technology parameter [6]. The minimum NF of the LNA is limited
by the NF_{min} of the technology. The maximum gain is limited by the G_{max} of the
technology. The IIP_3 is limited by the maximum current that the transistor can
afford. Within the circuit power consumption requirement, gain, linearity (IIP_3),
and noise (NF) are the triangle that trades off among each other. Because the total
power is limited in monolithic passive wireless powered sensor nodes, the power
is the dominant parameter. Thus, within this limited power, the strategy is trading
linearity for higher gain and lower NF.

7.3.4   Self-mixer

On-Off-Keying (OOK) is a widely used modulation scheme for low-power sensor
applications [7, 8, 9]. In the traditional wake-up radio, a low-biased diode detector
is used for envelop detection in OOK demodulation. However, the sensitivity of
such an approach is poor and requires an additional oscillator. An alternative is a
self-mixer, which is formed by an injection-locked oscillator (IJLO) and a passive
mixer, as shown in Fig. 7.4. By using an IJLO, the quadratic behavior of the WuRx
is replaced by a linear relation, and the sensitivity is improved [10]. Moreover the
frequency information of the incoming RF signal has been identified during the
locking process, thus a PLL is not required, and there is no power consumption for
the passive mixer as apposed to an active mixer. Because the power reduction is
from replacing a PLL by a IJLO with mixer, this architecture can achieve low power
consumption with high sensitivity.

The IJLO is composed of the input buffer stage, an LC core, and the output buffer.
The LC core part consists of a parallel resonant tank composed of an inductor and a
capacitor, and the negative g_m stage for compensating energy loss during each cycle.
Because the energy is stored inside the L or C, the output power of the IJLO is given
by [11]

$$P_{IJLO} = RC^2 \omega_C^2 V_{pk}^2 = \frac{R}{L^2 \omega_C^2} V_{pk}^2 \tag{7.6}$$

where ω_C is the oscillating frequency and V_{pk} is the peak voltage swing. The phase noise of the IJLO has direct impact on the receiver NF and the transmitted signal quality of the transmitter, so the phase noise of the IJLO should be low. The phase noise is inversely proportional to the power consumption. Based on Lesson's model [12], the phase noise power density, S_ϕ, is expressed as

$$S_\phi = \text{NEF} \frac{kT}{2P_{\text{sig}}} \frac{\omega_C^2}{Q^2 \Delta \omega^2} \tag{7.7}$$

where NEF is the noise excess factor, Q is the quality factor of the LC tank, P_{sig} is the AC signal power, $\Delta \omega$ is the offset frequency. Combining (7.6) and (7.7), P_{IJLO} can be expressed in (7.8).

$$P_{\text{IJLO}} = C \left(\frac{R}{L} \right)^3 \text{NEF} \frac{kT}{S_\phi} \frac{\omega_C^2}{Q^2 \Delta \omega^2} \tag{7.8}$$

The DC power consumption of the IJLO can be described related to the output power of the IJLO by power converting efficiency, η_{IJLO}, which is defined as

$$\eta_{\text{IJLO}} = \frac{P_{\text{IJLO}}}{P_{\text{DC\_IJLO}}} \tag{7.9}$$

Because a passive mixer is connected to the IJLO directly in the structure of the self-mixer, the buffer is not necessary for an IJLO, which will reduce the total power consumption. The mixer inside the self-mixer is a passive structure, so there is no DC power consumption. The output power of the self-mixer can be described as the power of the IJLO as in (7.10).

$$P_{\text{self-mixer}} = P_{\text{DC\_IJLO}} = \frac{C \left(\frac{R}{L} \right)^3 \text{NEF} \frac{kT}{S_\phi} \frac{\omega_C^2}{Q^2 \Delta \omega^2}}{\eta_{\text{IJLO}}} \tag{7.10}$$

7.3.5 System Limitations

In the analysis of the energy model, there are three main factors limiting the system performance. The first is the antenna gain. The second is the on-chip capacitor size. The third is the consumption loss due to the active components.

The antenna gain is limited in silicon, which is due to the lossy silicon substrate. An on-chip antenna array could solve this limitation, but the total area will limit the number of antennas. The on-chip capacitor is the energy storage unit. The total capacitance can be described as $C_{\text{total}} = C_{\text{density}} \times \text{Area}$, where C_{density} is the capacitor density per unit area. In CMOS technology, this is limited by the metal distance for the metal-oxide-metal capacitor or insulator gap for the metal-

insulator-metal capacitor. Enlarging area is an effective method to increase the total
capacitance value. The active components are the components that consume the
energy. The leakage current from the transistor determines the minimal required
power generated by the rectifier. The maximum breakdown voltage of the active
components limits the maximum power that can be stored. The loss from the *LC*
tank with the loss from the negative g_m transistor limits the minimal power for the
oscillator.

7.4   System Evaluation

The key parameters and power consumption of the RF front-end are shown in
Table 7.1a. To gauge the performance achievable with the proposed sensor node, we
consider a system with key parameters shown in Table 7.1b. The base-station has a
transmit power of 10 dBm and a transmit antenna gain of 20 dBi. The gain of the on-
chip antenna is assumed to be 0 dBi. The modulation is OOK with 2 GHz bandwidth
and the packet length is 20 bits. We also assume the sensor spends $T_{scav} = 1$ ms to
scavenge energy and 50% of the scavenged energy is used for receiving.

Figure 7.5 compares the energy received by the sensor node and the required
energy by the front-end for a 20-bit packet. We can see that the received energy is
sufficient for a communication distance of 10 cm, which is sufficient for RFID type
of applications. Moreover, we notice that due to the large input power available
after the receive antenna, the effective signal to noise ratio (SNR) is more than
sufficient (above 50 dB for the distance considered) for reliable reception. In some
applications, there is no need to consider the isolation between the antenna and
the IJLO, the LNA can be removed as well, which can further reduce the receiver
power. In that case, the communication distance is increased from 10 cm to 12 cm.
For the IJLO, there is a limitation on the minimum current consumption for starting
oscillation, which is around 2 mA for the VCO core part.

Table 7.1 RF front-end parameters and system parameters for the evaluation of the proposed sensor
node

(a) 60 GHz RF front-end parameters				
	Gain (dB)	NF (dB)	IIP$_3$ (dBm)	P_{DC} (mW)
LNA	16	5	−18	5
Mixer	−15	15	−22	
	f_{rf}	P_{out}	Sensitivity (dBm)	P_{DC} (mW)
IJLO	61	−22	−60	3
(b) System parameters				
Tx power	Ant. gain Tx		Ant. gain Rx	Modulation
10 dBm	20 dBi		0 dBi	OOK
BW	T_{scav}		Pac. Len	Rx %
2 GHz	1 ms		20 bits	50

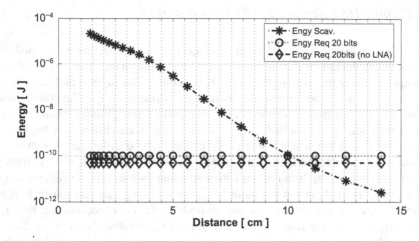

Fig. 7.5 System evaluation of proposed sensor node

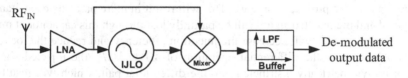

Fig. 7.6 The system block of the 60 GHz ULP OOK receiver

From this evaluation, it is noticed that the rectifier circuit needs 7 dBm input power to guarantee a 7% efficiency. This means that similar power is available at the receiver and the SNR is high. As a result, the sensor receiver is not SNR limited but power limited. More efforts should be spent on trading noise performance to lower power consumption.

7.5 Circuit Implementation for the 60 GHz Ultra-Low-Power Receiver

The simplified system architecture of a 60 GHz ultra-low-power receiver described in the previous section is re-drawn in Fig. 7.6. It is composed of the low noise amplifier (LNA), the injection-locked oscillator (IJLO), and the mixer. In this section the circuit design and implementation of the low power LNA, passive mixer, and IJLO will be presented. Based on the ultra-low-power radio design, a chip implemented in 65 nm CMOS technology is presented together with the measurement results.

7.5.1   60 GHz Injection-Locked Oscillator

In an injection-locked oscillator, a signal is injected into the oscillator core, and the core frequency locks to the incoming signal. Thereby the phase noise is reduced and an output signal at the locking frequency is generated. The input signal (e.g., from the antenna) can be injected into the oscillator in different ways. Generally, these can be divided into three methods: tail injection, direct voltage injection, and direct current injection, as shown in Fig. 7.7. These methods are different in frequency locking range and the sensitivity. Therefore they will be difference in mm-wave operation. The tail injection method is widely used in IJLO as well as in frequency dividers [13, 14, 15, 16]. It can be implemented either by injecting the voltage signal to the gate of the tail current transistor (M_{tail}) or by using a current mirror to feed the injection current directly to the source node of the cross-coupled negative-g_m transistors. Because the transistor M_{tail} is not connected directly to the oscillator core, it will not add extra parasitic capacitance to the oscillator tank and does not cause a loading effect. In order to generate a strong injection current, normally M_{tail} is large in order to provide a large g_m. This will result in more parasitic capacitance to ground and the injected signal will be partially lost through this capacitor at mm-wave frequency. This will decrease the injection efficiency and reduce the locking range. Also due to the differential operation of M_1 and M_2, the injected signal cannot be symmetrically distributed into the differential pair, which will result in an asymmetrical output signal. This could lead to a common-mode signal in the LO driver stage and reduce the gain of the mixer. The performance of direct voltage injection, Fig. 7.7b, is also deteriorated by the large injection transistor. In order to have small R_{on}, the injection transistor is large in the direct voltage injection method, and this will increase the parasitic capacitance to ground; decreasing the

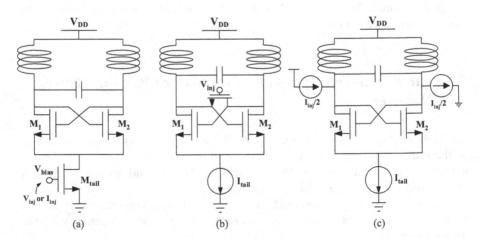

Fig. 7.7   IJLO topologies with 3 methods of injection: (**a**) tail injection can be both voltage and current injection, (**b**) direct voltage injection, (**c**) direct current injection

Fig. 7.8 Differential current
re-use input buffer

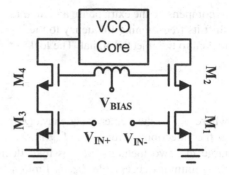

injection efficiency and also loading the LC tank seriously. The shunt inductor-peaking method [17] is adopted to compensate that parasitic capacitor, but it will occupy a large area and magnetic coupling to the LC tank can occur. In the direct current injection method, the current is injected symmetrically from both sides into the LC tank, and it can keep the current injection transistor small to decrease the loading effect on the LC tank. By optimizing the current density of the transistors, direct current injection can achieve good noise performance in order to increase the sensitivity of the IJLO and keep the high isolation between the LC tank and the input signal. In order to decrease parasitic loading and keep symmetric injection, the current-injection method is most suitable for IJLO in the mm-wave frequency range and is chosen for the implementation in this work as will be described in the next paragraphs [18].

The input buffer of the direct current injection can be implemented as a differential pair with current-reuse cascade topology, as shown in Fig. 7.8. By using the cascode topology, the common source (CS) transistors (M_1, M_3), and common gate (CG) transistors (M_2, M_4) can separate the design targets of low noise and high gain. The input CS transistor is biased at low noise current density, $120 \, \mu A/\mu m$, to provide good sensitivity, and the CG transistor is biased at high gain current density. Furthermore two stacked transistors can improve the isolation between the injected signal port and the LC tank. By using an inductor at the gate of the CG transistor, this topology can take advantage of the current-reuse topology [3]. The drain current of M_2 can be described as:

$$i_{d2} = g_{m2}v_{gs2} = \frac{\omega_{T2}}{\omega} \frac{i_{d1}}{j} \qquad (7.11)$$

where i_{d1} and i_{d2} are the drain currents of M_1 and M_2, respectively, g_{m2} is the trans-conductance of M_2, v_{gs2} is the gate-source voltage of M_2, ω is the operating angular frequency, and ω_{T2} is the angular cut-off frequency of M_2. In other words, the cascode stage is then converted into a cascaded topology. The current gain is theoretically increased while the DC current is reused by these two stages and stays unchanged.

When an oscillator is under injection, e.g. at its resonator, a phase shift is created between its oscillation signal and the resonator path [19]. As shown in [19], in order to compensate the extra phase and maintain 360° loop phase shift, the oscillator will shift its free-running frequency to the injected one. In other words, the oscillator is locked to the injected signal. The locking range ω_L is calculated in (7.12)

$$\omega_L = \frac{\omega_o}{Q} \times \frac{I_{\text{inj}}}{I_{\text{osc}}} \tag{7.12}$$

where ω_o is the center frequency, Q is the quality factor of the resonator, I_{inj} is the injection current, and I_{osc} is the oscillating current. In order to obtain a large ω_L, two methods are possible: decreasing I_{osc} or increasing I_{inj}. However, the minimum acceptable I_{osc} is limited by the IJLO core, i.e. it must be large enough to maintain the oscillation condition. Consequently, the trade-off between the IJLO core injection current sensitivity (I_{inj}) and the locking range inherently limits the overall performance and makes the IJLO hardly suitable to be used in the RF-path of a front-end, where the signal is normally weak while its bandwidth is relatively large especially in the mm-wave bands. The solution is to use the frequency sweeping acquisition method, i.e. sweeping the center frequency of the IJLO with certain resolution until the IJLO is locked with the injection signal. In this way, the physical locking range of the IJLO does not need to be large while the overall achievable locking range is extended significantly. However, the system latency is also increased. Therefore, the sweeping resolution must be optimized to make sure the sensitivity requirement is achieved while the sweeping process is sufficiently fast [10].

The schematic of the implemented injection-locked oscillator is shown in Fig. 7.9. The oscillator LC tank is implemented by a center-tapped inductor, and two cross-coupled common-source transistors to provide negative g_m. The transistor is biased at a low current point with large f_{max}, and at the same time still provides enough negative g_m to oscillate. Because it is the symmetrical direct current injection method, the input buffer adapts current-reuse technology to increase the current-gain. The current-reuse method provides high gain while keeping the input buffer current consumption small. From simulation, the sensitivity is -60 dBm with 6 mW DC power consumption.

7.5.2   60 GHz Low Power Differential LNA

In order to further improve the sensitivity of the IJLO based demodulation, a low noise amplifier is required in front. The LNA is the first stage and located directly after the antenna matching network. It provides RF gain, noise matching in order to ensure low noise figure [20]. Because of the differential structure of the input buffer of the IJLO, the low noise amplifier in this architecture is implemented with a differential output.

The LNA is designed in 65 nm CMOS technology. In this technology, there are two optimized current densities. One current density is optimized for low noise performance, the other current density is optimized for speed. In this technology,

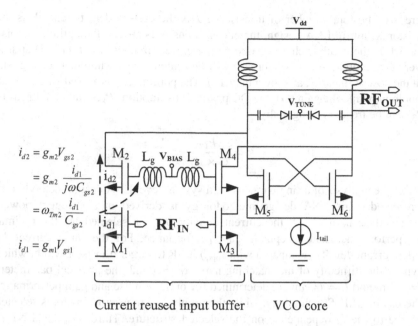

$$i_{d2} = g_{m2}V_{gs2}$$

$$= g_{m2}\frac{i_{d1}}{j\omega C_{gs2}}$$

$$= \omega_{Tm2}\frac{i_{d1}}{C_{gs2}}$$

$$i_{d1} = g_{m1}V_{gs1}$$

Current reused input buffer   VCO core

Fig. 7.9 Schematic of the injection-locked oscillator

the typical value of minimum noise factor in the NMOS transistor is $J_{opt_{noise}} = 0.15\,\text{mA}/\mu m$. And the value for maximum frequency is $J_{opt_{max}} = 0.2\,\text{mA}/\mu m$. Those two values remain constant for different combinations of the finger width and the finger number. For the transistor size, there are two parameters, length (L) and width (W). Normally, the L is chosen to be the minimum value to optimize performance. Because the gate is made of the poly-silicon material with resistivity of $15\,\Omega/\square$, the longer the gate, the larger the gate resistance. The gate resistance can be described as [21]

$$R_{gate} = \frac{R_{gsq}}{3}\frac{W_f}{N_f L_g} + \frac{R_{cont}}{N_{cont}N_f} + \frac{R_{gsq}}{N_f}\frac{l_{access}}{L_g} \tag{7.13}$$

where R_{gsq} is the gate sheet resistance. W_f is the finger width. N_f is the number of fingers. L_g is the gate length. L_{access} is the distance between gate contact and active area. R_{cont} is the metal-poly contact resistance. N_{cont} is the number of contacts per gate finger.

Based on the bias current density, the power consumption of the transistor can be expressed as the product of the biasing current and the voltage across it, V_{ds}. Because the drain of the transistor is connected to the supply voltage, V_{DD}, the power consumption of the LNA can be expressed as

$$P_{DC} = V_{DD}\left(\sum_1^n I_{bias}\right) = V_{DD}\left(\sum_1^n J_{bias}W\right) \tag{7.14}$$

where n is the total number of transistors, J is the current density, and W is the transistor width. In LNA design, the current density is around the optimized noise point, while the width is chosen based on the g_m requirement. In order to keep low power consumption, the transistor size will be scaled down, while that scaling will be at the cost of linearity, as shown in (7.15). The parameter k is a fixed value for the technology, while the gain (G), the DC power consumption (P_{DC}), and the linearity (IIP_3) can be treated orthogonally.

$$k = \frac{P_{DC}}{G \times IIP_3} \tag{7.15}$$

After the analysis of a single transistor, a system level LNA design methodology is presented. This LNA design methodology is derived in a stringent power-constrained scenario. First, the current density, J_{opt}, is selected for the optimal noise performance and acceptable gain performance. For the same current I_d, smaller current density (compared with J_{opt_f}) leads to larger transistor width, which alleviates the difficulty of the matching network. Second, the physical parameters $w_p = 1\,\mu m$ and $L = 60\,nm$ are determined for optimal noise and gain performance in the 65 nm CMOS technology, while the number of transistor fingers is selected later due to the I_d dependence on the selected structure. Third, G_{max} and NF_{min} are obtained from simulation results. Fourth, the minimal number of transistors (N_t) is fixed by the total gain of the LNA and G_{max} of each transistor. Fifth, the optimal structure including stage numbers (N_s) and transistor numbers per stage is determined [22].

The LNA is implemented in the differential way, composed of cascaded two-stage cascode structure, as shown in Fig. 7.10. The input of the LNA is matched to $50\,\Omega$. The degeneration inductor provides the stability and broadband power and noise matching. The inter-stage inductor between the common source and

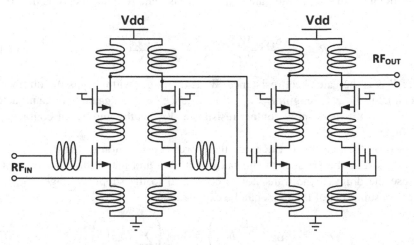

Fig. 7.10 The schematic of 60 GHz LNA

the common drain removes the parasitic capacitors improving the stability and achieving the inter-stage matching. During the implementation, the load inductor can be implemented by a centrally tapped inductor to save area. The inter-stage matching can be implemented as a transmission line inductor. The simulated LNA shows 10 dB gain, and 5 dB NF at 60 GHz, with 9 mW power consumption.

7.5.3 60 GHz Passive Mixer

In ultra-lower power radios, a passive mixer is a good choice for its low power consumption and good linearity. As there is no DC current, there is no flicker noise if both switch pairs do not switch on at the same time. The loss generated by the passive mixer can be compensated at the IF frequency by a low frequency gain stage. A differential double-balanced mixer structure can enhance the LO-IF and RF-IF isolation. The differential structure also eliminates the problem of the output common-mode. Also the symmetrical structure needs to be taken into account in the layout, including the symmetrical effective distance for the signal length from the LO to the mixer. The structure and conceptual model of the double-balanced passive mixer is shown in Fig. 7.11. The mixer core transistors are modeled as an ideal switch with an on resistance, R_{on} and off impedance, Z_{off}. The off impedance can be estimated as $1/j\omega C_{off}$ in which C_{off} is the combination of the gate-drain, gate-source, and source-drain parasitic capacitances of the transistor in the cut-off mode.

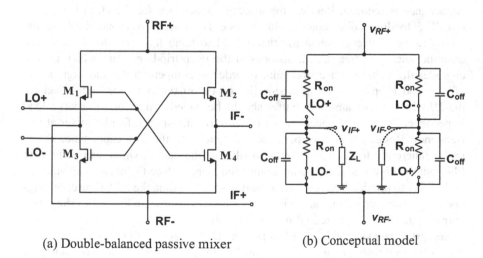

(a) Double-balanced passive mixer   (b) Conceptual model

Fig. 7.11 (a) Schematic and (b) conceptual model of a double-balanced passive mixer

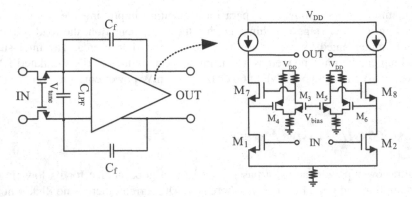

Fig. 7.12 Output buffer of the mixer

The voltage conversion gain of such a double-balanced passive mixer can be described as [23]:

$$G_{c,v} = 20\log\left(\frac{2}{\pi}\left(\frac{Z_{\text{off}} - R_{\text{on}}}{\left(1 + \frac{R_{\text{on}}}{Z_L}\right)\left(1 + \frac{Z_{\text{off}}}{Z_L}\right)}\right)\right)$$
(7.16)

From (7.16), it can be seen that three methods can be used to improve the conversion gain, i.e. increasing Z_{off}, decreasing R_{on}, and increasing Z_L. Increasing Z_{off} means that smaller transistors should be chosen so that the gate-source and gate-drain capacitance is reduced. Besides, the loading capacitance for the IJLO is reduced too. The loading effect could influence the IJLO frequency and decrease the tuning range. However, small transistors lead to large R_{on} (i.e., the drain-source on-impedance r_{ds} when the transistor is in the deep-triode region), which in turn degrades the voltage conversion gain. In order to compensate for the degradation of $G_{c,v}$ and keep the small transistor size of the mixer as the optimum load for the IJLO, an output amplifier buffer should be added, which offers a large load impedance, i.e. the gate capacitance of small-size transistors, for the mixer at low frequency. Besides, the series parasitic resistance and the gate capacitance of the buffer inherently form an R-C low pass filter at the output, which can be used to filter out the high-frequency output components and reduce the noise bandwidth.

In order to provide large gain after the down-conversion, the folded cascade gain-boosting structured output buffer is implemented, as shown in Fig. 7.12. The mixer output is fed to the differential pair M_1 and M_2. M_3 and M_4, M_5 and M_6 form a folded extra gain stage. M_7 and M_8 form the output stage. In this way, the amplifier gain is boosted without requiring additional voltage headroom. Instead of tuning the feedback capacitors C_f, the bandwidth tuning is realized by tuning the gate voltages of the NMOS resistors by V_{tune}. The buffer takes 1.4 mA from 1 V power supply, the 3 dB bandwidth is 1.3 GHz, and voltage gain is 17.8 dB.

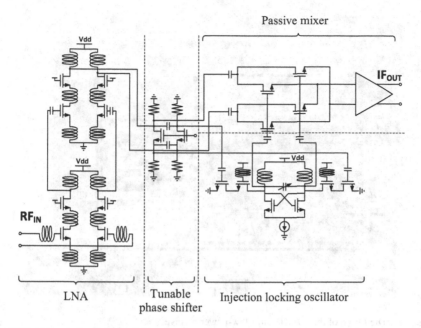

Fig. 7.13 Schematic of the 60 GHz ultra-low-power receiver

7.5.4 60 GHz Ultra-Low-Power OOK Receiver

The schematic of the 60 GHz receiver is shown in Fig. 7.13. At the input the RF signal is amplified by the LNA, the signal is divided into two paths. One path to the IJLO, which is used to generate the clock signal, while the other is fed to the mixer. Because of the phase difference between the two signal paths, a phase shifter is inserted in front of the mixer. The output of the mixer, the IF signal, is fed to the filter, to provide enough gain and provide a low pass filter function.

7.5.5 60 GHz Ultra-Low-Power OOK Receiver Measurement

The chip was fabricated in 65 nm CMOS technology, and the chip photo is shown in Fig. 7.14. The principle of the self-mixer based receiver is by self-mixing in a locked IJLO. Thus, there are two situations, the first is when the IJLO is not locked, and the second is the locked situation. Figure 7.15 shows the measured IF spectrum under the situation that IJLO is not locked to the input RF signal. Figure 7.16 is the measured output signal spectrum with the situation that IJLO is locked to input RF signal. Comparing Figs. 7.15 to 7.16, there are two frequency components while there is only one frequency component in the locked siltation (Fig. 7.16), and IF power is −37 dBm, while the unlocked situation only −50 dBm.

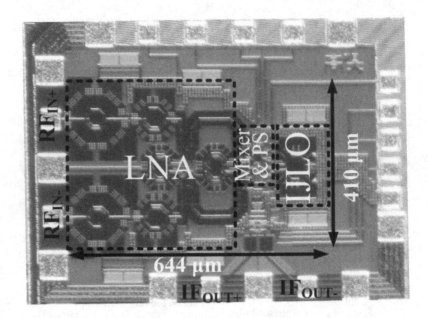

Fig. 7.14   Die photo of the 60 GHz ultra-low-power receiver

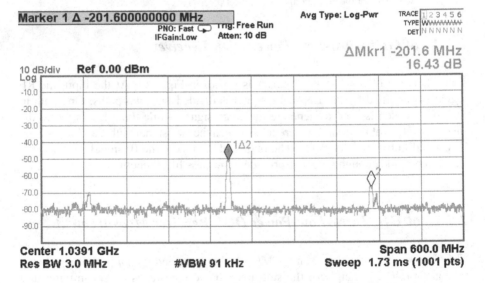

Fig. 7.15   The measured spectrum of the down-converting in the unlocked situation

The input carrier signal and modulated data signal are located at the frequency of f_{RF} and $f_{RF} + f_{sig}$. And the free-running LO signal is f_{LO}. In the unlocked situation, the input carrier signal and modulated data signal are down-converted by the free-running LO signal in the self-mixer. After the down-conversion, the original input

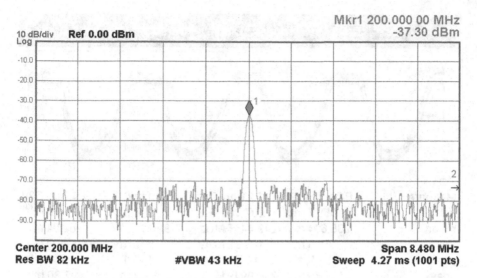

Fig. 7.16 Measured spectrum of the down-converting in the locked situation, the output is IF frequency

carrier signal and modulated data signal are mixed down to $f_1 = f_{RF} - f_{LO}$ and $f_2 = f_{RF} + f_{sig} - f_{LO}$. In the locked situation, the IJLO is locked to the incoming signal, so $f_{LO} = f_{RF}$. After the down-conversion, the $f_1 = f_{RF} - f_{LO} = 0$, and $f_2 = f_{RF} + f_{sig} - f_{LO} = f_{sig}$. Thus, only one signal located at the f_{sig} will be detected. Figure 7.16 shows the output 200 MHz signal is de-modulated. The time-domain de-modulated signal is shown in Fig. 7.17. The total power consumption of this receiver is 16.4 mA with 1 V supply.

The summary and comparison of this work to the state-of-the-art receivers is listed in Table 7.2. This work demonstrates the concept of IJLO based self-mixer. It can achieve low power consumption. Compared with other work, the total DC power consumption of this work is low and has the potential to be intergraded with the fully passive sensor node with wireless power transfer function. The total chip area is smaller than [24, 25, 26], which can provide more area for the on-chip capacitor for energy storage.

7.6 Conclusion

In this chapter, the radio-triggered passive receiver is introduced. Energy models for key components, including the matching network, the rectifier, the LNA, and the self-mixer are developed for a wireless powered monolithic sensor receiver using the radio-triggered passive receiver architecture. Analysis of the efficiency of wireless power transfer and the frequency choice reveals that the 60 GHz band is suitable for this application. Using the developed energy models, a system evaluation of

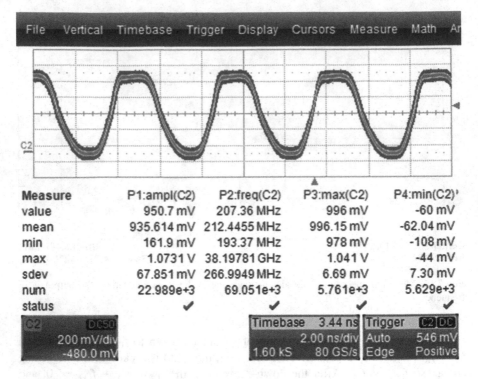

Fig. 7.17 Measured data stream in the locked situation

Table 7.2 Performance comparisons for the receivers

	[24]	[25]	[26]	This work
CMOS technology (nm)	90	65	90	65
Frequency (GHz)	60	60	62	60
Modulation	OOK	BPSK	OOK	OOK
P_{diss} (mW)	103	151	49	16.4
Area (mm^2)	0.68	1	0.72	0.26

the performance of the sensor receiver is performed and it shows that reliable communication can be achieved over a distance of 10 cm, which is sufficient for RFID type of applications. Furthermore, it is identified that the receiver is power limited rather than SNR limited. This suggests to further improve the system by focusing on trading noise performance to lower power consumption. Based on the system evaluation, an injection-locked oscillator based mm-wave ultra-low-power receiver is designed. The ultra-low-power receiver is implemented in 65 nm CMOS technology. It demonstrates the concept of the IJLO based self-mixer. The total power consumption of this receiver is 16.4 mA with 1 V supply. The area of this receiver is much smaller than other implementations, Table 7.2, which can free up more area in the passive sensor node for energy storage.

References

1. Y. Wu, J. Linnartz, H. Gao, P. Baltus, J. Bergmans, System study of a 60 GHz wireless-powered monolithic sensor system, in *2011 8th International Conference on Information, Communications and Signal Processing (ICICS)*, December 2011, pp. 1–5
2. S. Drago, D. Leenaerts, F. Sebastiano, L. Breems, K. Makinwa, B. Nauta, A 2.4 GHz 830 pJ/bit duty-cycled wake-up receiver with −82 dBm sensitivity for crystal-less wireless sensor nodes, in *2010 IEEE International Solid-State Circuits Conference Digest of Technical Papers (ISSCC)*, February 2010, pp. 224–225
3. X. Huang, S. Rampu, X. Wang, G. Dolmans, H. de Groot, A 2.4 GHz/915 MHz 51 μW wake-up receiver with offset and noise suppression, in *2010 IEEE International Solid-State Circuits Conference Digest of Technical Papers (ISSCC)*, February 2010, pp. 222–223
4. J. Ryu, M. Kim, J. Lee, B.-S. Kim, M.-Q. Lee, S. Nam, Low power OOK transmitter for wireless capsule endoscope, in *IEEE/MTT-S International Microwave Symposium, 2007*, June 2007, pp. 855–858
5. J. Liu, C. Li, L. Chen, Y. Xiao, J. Wang, H. Liao, R. Huang, An ultra-low power 400 MHz OOK transceiver for medical implanted applications, in *2011 Proceedings of the ESSCIRC (ESSCIRC)*, September 2011, pp. 175–178
6. P. Baltus, Minimum power design of RF front ends, Ph.D. dissertation, Eindhoven University of Technology, 2004
7. N. Roberts, D. Wentzloff, A 98 nW wake-up radio for wireless body area networks, in *2012 IEEE Radio Frequency Integrated Circuits Symposium (RFIC)*, June 2012, pp. 373–376
8. S. Oh, N. Roberts, D. Wentzloff, A 116 nW multi-band wake-up receiver with 31-bit correlator and interference rejection, in *2013 IEEE Custom Integrated Circuits Conference (CICC)*, September 2013, pp. 1–4
9. T. Wada, M. Ikebe, E. Sano, 60-GHz, 9-μW wake-up receiver for short-range wireless communications, in *2013 Proceedings of the ESSCIRC (ESSCIRC)*, September 2013, pp. 383–386
10. X. Li, P. Baltus, P. van Zeijl, D. Milosevic, A. van Roermund, A 73 to 83 GHz, 9-mW injection-locked oscillator in 65-nm CMOS technology, in *2011 IEEE 11th Topical Meeting on Silicon Monolithic Integrated Circuits in RF Systems (SiRF)*, January 2011, pp. 5–8
11. T.H. Lee, *The Design of Low Noise Oscillators* (Springer Science Business Media, Boston, 1999)
12. B. Razavi, *RF Microelectronics* 2nd edn. (Prentice Hall, Upper Saddle River, 2011)
13. M. Kennedy, H. Mo, X. Dong, Experimental characterization of Arnold tongues in injection-locked CMOS LC frequency dividers with tail and direct injection, in *2011 20th European Conference on Circuit Theory and Design (ECCTD)*, August 2011, pp. 484–487
14. Y. Liu, Z. Li, H. Gao, Q. Li, Z. Wang, A novel complementary push-push frequency doubler with negative resistor conversion gain enhancement. IEICE Electron. Express **14**, 20170674 (2017)
15. K.W. Li, L. Leung, K.N. Leung, Low power injection locked oscillators for MICS standard, in *2009. IEEE Biomedical Circuits and Systems Conference, 2009. BioCAS*, November 2009, pp. 1–4
16. A. Musa, R. Murakami, T. Sato, W. Chaivipas, K. Okada, A. Matsuzawa, A low phase noise quadrature injection locked frequency synthesizer for MM-wave applications. IEEE J. Solid-State Circuits **46**(11), 2635–2649 (2011)
17. J.-C. Chien, L.-H. Lu, Design of wide-tuning-range millimeter-wave CMOS VCO with a standing-wave architecture. IEEE J. Solid-State Circuits **42**(9), 1942–1952 (2007)
18. H. Gao, M.K. Matters-Kammerer, X. Li, D. Milosevic, A. van Roermund, P.G.M. Baltus, A 60-GHz injection locked oscillator for self-demodulation ultra-low power radio in 65-nm CMOS, in *2014 IEEE 21st Symposium on Communications and Vehicular Technology in the Benelux (SCVT)*, November 2014, pp. 90–93
19. B. Razavi, A study of injection locking and pulling in oscillators. IEEE J. Solid-State Circuits **39**(9), 1415–1424 (2004)

20. Z. Chen, H. Gao, D.M.W. Leenaerts, D. Milosevic, P.G.M. Baltus, A 16–43 GHz low-noise amplifer with 2.5–4.0 dB noise figure, in *2016 IEEE Asian Solid-State Circuits Conference (A-SSCC)*, November 2016, pp. 349–352
21. C.P. John, W. Roger, *Radio Frequency Integrated Circuit Design*, 2nd edn. (Artech House, Norwood, 2010)
22. Z. Chen, H. Gao, R. van Dommele, D. Milosevic, P.G.M. Baltus, Design consideration of 60 GHz low power low-noise amplifier in 65 nm CMOS, in *2016 Symposium on Communications and Vehicular Technologies (SCVT)*, November 2016, pp. 1–4
23. K. Komoni, S. Sonkusale, G. Dawe, Fundamental performance limits and scaling of a CMOS passive double-balanced mixer, in *2008 Joint 6th International IEEE Northeast Workshop on Circuits and Systems and TAISA Conference, 2008. NEWCAS-TAISA 2008*, June 2008, pp. 297–300
24. J. Lee, Y. Chen, Y. Huang, A low-power low-cost fully-integrated 60-GHz transceiver system with OOK modulation and on-board antenna assembly. IEEE J. Solid-State Circuits **45**(2), pp. 264–275 (2010)
25. A. Tomkins, R. Aroca, T. Yamamoto, S. Nicolson, Y. Doi, S. Voinigescu, A zero-IF 60 GHz 65 nm CMOS transceiver with direct BPSK modulation demonstrating up to 6 Gb/s data rates over a 2 m wireless link. IEEE J. Solid-State Circuits **44**(8), 2085–2099 (2009)
26. A. Oncu, M. Fujishima, 49 mW 5 Gbit/s CMOS receiver for 60 GHz impulse radio. Electron. Lett. **45**(17), 889–890 (2009)

Chapter 8
mm-Wave Front-End Design for Phased-Array Systems

Abstract In the previous chapters, a monolithic mm-wave sensor network was introduced. An on-chip wireless power receiver with an ultra-low-power receiver and transmitter front-end was presented. In this chapter, the base-station for monolithic sensor networks with phased-array architecture is analyzed and the key circuits are developed. By using a phased-array architecture, the base-station can achieve better sensitivity for the receiver part, and can also increase the transferred power density at the sensor node location for the transmitter part.

8.1 Introduction

In ultra-low-power systems, the transmitted power is limited by the available power consumption of the circuits and limited power added efficiency of the PA in the mm-wave range. In monolithic passive sensor nodes, the power is wirelessly transferred from a base-station to the sensor nodes, rectified in the sensor nodes and stored as electric energy on a capacitor. A high transferred power density leads to an increased efficiency of the rectification process and therefore to faster charging or longer range. A phased-array system architecture in the base-station increases the antenna gain and thereby increases the power density at the sensor node [1]. As a result of the higher efficiency rectification process, the charging distance between the sensor node and the base-station can be increased. At the same time, the link-budget for wireless communication is also improved, leading to a superior system sensitivity. So for charging and communication, phased-array systems are a key advantage. The system architecture is shown in Fig. 8.1. In this chapter the building block design of the base-station receiver (Rx) for the phased-array system is described.

8.2 Link Budget of the 60 GHz Sensor Network

For a typical system architecture, the link budget of the signal transferred from a sensor node to the base-station is shown in Table 8.1. The output power of the

© Springer International Publishing AG 2018
H. Gao et al., *Batteryless mm-Wave Wireless Sensors*, Analog Circuits and Signal Processing, https://doi.org/10.1007/978-3-319-72980-0_8

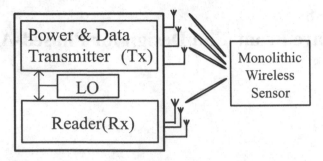

Fig. 8.1 System illustration of the phased-array base-station

Table 8.1 Link budget of a typical 60 GHz wireless sensor network

Component	Contribution	Signal power
Tx power (sensor nodes)	−30 dBm	−30 dBm signal
Tx antenna gain (sensor nodes)	+2 dBi	
Path loss over 0.5m	−62 dB	
Rx Array Antenna Gain (12 element)	$(+16) + (12) = 28$ dBi	−62 dBm signal
Background noise	−174 dBm/Hz	−174 dBm/Hz noise
Noise BW (1 GHz)	+90 dB	−84 dBm noise
SNR		$(-62) - (-84) = 22$ dB
Required SNR		10 dB (i.e., for OOK)
System margin with Rx NF		$22 - 10 = 12$ dB

transmitter (Tx) inside the monolithic wireless sensor nodes is −30 dBm. The on-chip antenna of the sensor node provides 2 dBi gain. According to Friis equation [2], the received power is −62 dBm assuming the base-station is half a meter away from the sensor node. The base-station provides 28 dBi receiver antenna gain. The system noise floor is −84 dBm under the condition of 1 GHz bandwidth. The required SNR for the OOK modulation with 1 Gbps is 10 dB. The system margin including the Rx NF is 12 dB. The calculation assumes 28 dBi Rx antenna gain. This 28 dBi antenna gain can be achieved by a phased-array system architecture with 12 antenna elements. Each Rx antenna has a gain of 16 dBi.

8.3   Phased-Array Architecture

8.3.1   Advantages of a Phased-Array Receiver Architecture

The 7 GHz unlicensed bandwidth at the 60 GHz ISM band [3] makes it attractive for high data rate applications. The decreasing antenna size with frequency leads on the one hand to tiny sensor nodes, but on the other hand to a high free space path loss of, e.g., 68 dB for 1 m distance. Compared to this, atmospheric attenuation (19 dB/km) is less problematic for the envisioned in-door sensor nodes.

In a phased-array transmitter, the signals transmitted by the array of antennas can be added up coherently in certain directions in space through power combining. In comparison to a single-antenna transmitter that transmits an output power of P_0, each path of the phased-array transmitter can transmit an output power of P_0/N and keep the sum of the output power equal to P_0. The equivalent isotropic radiated power (EIRP) of the phased-array transmitter will be $P_0 \times N^2$, which is increased by $10 \times \log_{10} N$ dB in comparison to a single-antenna transmitter.

In a phased-array receiver, the signal from the desired direction adds up coherently, while signals from all other direction will be partially or even fully canceled. The noise factor is defined as the ratio of the SNR at the input to the SNR at the output. The noise factor F_{RX} of an N-element phased-array receiver can be expressed as [4]:

$$F_{\text{RX}} = \frac{N \times \text{SNR}_{\text{in,single-element}}}{\text{SNR}_{\text{out,array}}} = NF_{1:N} \tag{8.1}$$

where $nF_{1:N}$ refers to the noise factor of an N-element phased-array receiver (RX) when a single element input is taken with respect to an N-element output, expressed as

$$F_{1:N} = \frac{\text{SNR}_{\text{in,single-element}}}{\text{SNR}_{\text{out,array}}} \tag{8.2}$$

Assuming that the noise is dominated by the noise of the front-end (FE) of each path, and ignoring other receiver components,

$$F_{1:N} = \frac{F_{\text{FE}}}{N} \tag{8.3}$$

Therefore the SNR at the output is N times the SNR at the output of each front-end. And the noise factor of a phased-array RX is close to the noise factor of each front-end. Assuming a parameter η ($\eta \leq 1$) denoting effects such as antenna coupling and influence of other RX components, the noise factor of a receiver array could be expressed by (8.4), while the receiver array output SNR can be expressed by (8.5).

$$F_{\text{RX}} = \frac{N \times \text{SNR}_{\text{in,single-element}}}{N \times \eta \times \text{SNR}_{\text{out,FE}}} = \frac{F_{\text{FE}}}{\eta} \tag{8.4}$$

$$\text{SNR}_{\text{out,array}} = N \times \eta \times \text{SNR}_{\text{out,FE}} \tag{8.5}$$

The ability of SNR enhancement enables the phased-array RX to achieve better sensitivity compared to a single element RX.

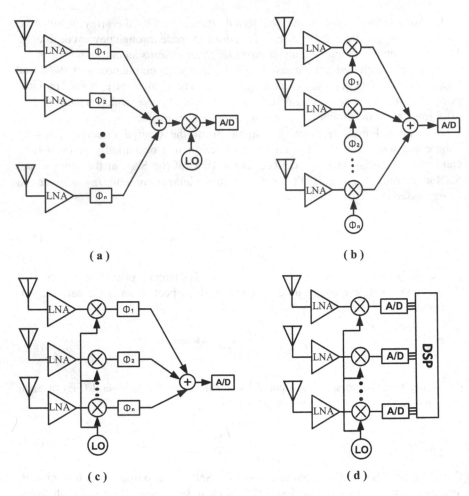

Fig. 8.2 Phased-array architectures with different phase shifting schemes. (**a**) RF phase shifting; (**b**) LO phase shifting; (**c**) IF phase shifting; (**d**) Digital baseband phase shifting

8.3.2   Signal Path Phase Shifting

There are four locations suitable for the phase shifter to achieve the phase shifting function in the receiver. Those are in the RF path, in the LO path, in the IF path, or in the digital domain, as shown in Fig. 8.2.

The RF phase shifter based architecture is shown in Fig. 8.2a. In this architecture, the signal combining and phase shifting are carried out in the RF path [5, 6, 4, 7]. The LNA and the phase shifter compose an RF front-end for each antenna path, and other radio blocks are shared resulting in reduced area and power consumption. Additionally since interference signals can be canceled after power combining, both noise figure and linearity requirements of following blocks can be relaxed, allowing compensation for other system requirements such as power consumption. However the phase shifting at RF introduces losses in the RF signals which affects the receiver sensitivity and the quality of received signal.

The LO phase shifter based architecture is shown in Fig. 8.2b. In this architecture, phase shifting is implemented in the LO path and power is combined at IF. Phase shifting in the LO path [8, 9] is advantageous in that the loss of the phase shifter does not directly deteriorate the receiver sensitivity. Besides, the power is combined at IF which is easier to implement. It requires duplication of more circuits and the distribution of the LO signal at mm-wave frequencies is challenging. Since interferers are not canceled before the mixer, the mixer has higher dynamic range requirement and linearity.

The IF phase shifter based architecture is shown in Fig. 8.2c. In this architecture, the phase shifting and power combining are carried out in the IF path, which requires a relatively broadband phase shifter (compared to the center frequency). The value of the passive components such as inductors and capacitors for a certain phase shift is inversely proportional to the carrier frequency. Since the values of integrated passive components are directly related to their physical size, passive phase shifters at IF consume a larger area compared to the ones at RF.

The architecture with phase shifting implemented in the digital domain is shown in Fig. 8.2d. In this architecture, the delay is implemented in the digital baseband. Phase shifting in the digital domain has advantages of flexibility and accuracy. However it also requires duplication of the RF/IF signal paths, including mixers, analog-to-digital converter (ADC), resulting in a large area and high power consumption. Besides, interference signals are canceled out only after phase shifting, therefore all circuit blocks need to have a large dynamic range to process these interferers without degrading the signal of interest. This increases the complexity of the RF and ADC blocks and their power consumption.

Based on the analysis of all four phase shifting structures, the RF path phase shifter architecture is a compact solution [10]. The challenge for the 60 GHz phase shifter is to achieve low loss and large bandwidth. To compensate for the phase shifter insertion loss, a variable-gain-amplifier (VGA) is used to reduce its deteriorating effect on the receiver sensitivity [11]. As only one mixer is needed for the entire array, the core circuitry of the receiver (front-end, up to the mixer) can be reused for each path of the array, without the need to add additional mixers. Thus makes it simple to extend to multiple antennas. Therefore the main challenge is to design and implement an LNA with low noise figure, a phase shifter with good phase accuracy, low loss and small phase/loss deviation and a VGA with different gain compensation at various settings of the phase shifter.

8.3.3 RF Front-End and Specification

Figure 8.3 shows a block diagram of a receiver front-end, which consists of a low-noise amplifier (LNA), a phase shifter, and a variable gain amplifier (VGA). The phase shifter has the function to change the delay in the RF path. This enables

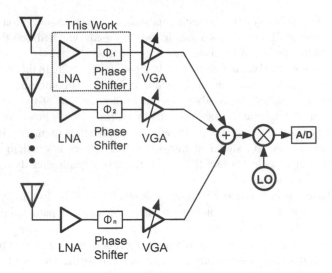

Fig. 8.3 Block diagram of a receiver front-end for an RF phased-array system

electronic beam scanning. The VGA has the function to provide adjustable gain for different phase settings. This can make the total RF gain more flat in the front-end. The LNA is located in front of the phase shifter and VGA to increase the SNR. From the link budget calculation, the system margin for the RX NF is only 12 dB, Sect. 8.2.

In the analog domain, phase shifters could be implemented using structures of switch-type [12, 13], vector-sum [14, 15], or reflection-type [16, 17]. Vector-sum phase shifters provide the required insertion phase by adjusting the weighting of the quadrature-phased signals. Reflection-type phase shifters are popular for continuous phase tuning, controlled by tunable varactor loads. However phase shifters using these two techniques require a high resolution DAC to provide the specific phase shift. These DACs will consume more power to convert digital control bits to analog signals. Also, power consumption for high speed requirement is high for those DACs. Therefore a switch-type digitally controlled phase shifter is preferred over the others. Furthermore, it is more robust than the other options to on-chip interference [18].

The switch-type phase shifters are typically implemented with 4 control bits [19, 20] or 5 control bits [21, 22] to satisfy the requirement of angular resolution. Phase delay resolution depends on array size. A 5-bit digitally controlled phase shifter has a phase delay resolution of $11.25°$ (a maximum phase error of $\pm 5.625°$). Figure 8.4 presents the array factor of a 16-element linear phased array with 4-bit and 5-bit phase shifters, with an antenna spacing of $d = \lambda/2$. It can be seen that a 4-bit phase shifter is sufficient for all angles at close to peak array gain. Although a 4-bit phase shifter is sufficient for satellite communication and radar systems [21], it is not enough for high data rate (multi-Gbps) systems that use high order modulation

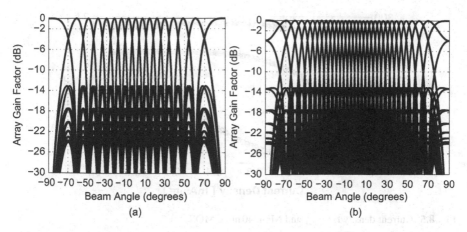

Fig. 8.4 Array factor of a 16-element phased array with (**a**) 4-bit phase shifter, (**b**) 5-bit phase shifter

scheme such as 16-QAM or 64-QAM and large bandwidth. A 5-bit phase shifter has a higher phase delay resolution to help reduce the error vector magnitude (EVM). It is also clear that the loss at the scanning direction due to the 4-bit phase resolution is less than 2 dB, while the loss could be reduced to less than 0.5 dB if 5-bit phase shifter is used. Thus, 5-bit phase shifter with 11.25° step is chosen in this design.

8.4   60 GHz LNA

This section presents a 60 GHz low noise amplifier realized in a 40 nm CMOS technology. The design incorporates additional noise matching between the common-source stage and the common-gate stage to reduce the noise impact by the latter stage. The measured noise parameters indicates an NF of 3.8 dB at 60 GHz. The achieved 3 dB-power gain bandwidth is 13 GHz, with peak G_t of 15 dB at 55 GHz, and of 12.5 dB G_t at 60 GHz. The total power consumption is 20.4 mW.

8.4.1   Technology

The LNA is realized in a 40 nm CMOS technology. In this technology, the RF NMOS transistor has the peak f_{max} of 290 GHz at a current density of 0.37 mA/µm, and the NF_{min} at a current density of 0.18 mA/µm, as shown in Fig. 8.5. 60 GHz is at 1/5 of f_{max}, which allows for a safe margin for RF design. The back-end structure of this technology is 1 poly layer and 8 metal layers with top Cu metal of 1.7 µm thickness, and the resistivity of the substrate is 10 Ω-cm [23].

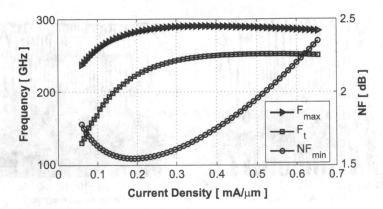

Fig. 8.5 Current density for F_{max} and NF in 40 nm CMOS

8.4.2   Topology Selection

As the first block of a receiver, an LNA is expected to amplify the signal with a low NF, while fulfilling the specified requirements on gain, input matching, bandwidth, and linearity. Due to the limited design freedom, it is difficult to achieve all these targets in a single-stage LNA at 60 GHz in CMOS technology. Therefore, the proposed LNA employs a two-stage architecture. The first stage provides a low noise figure and an input termination to 50 Ω, while the second stage is mainly optimized to provide a high available power gain. Since the first stage of the design dominates the overall noise performance of the LNA, it is crucial to choose a topology which gives the best performance at 60 GHz for the given technology. Figure 8.6 shows three topology candidates that will be discussed and compared in detail.

1. Common gate (CG) structure
   One of the superior features of the CG structure, Fig. 8.6a, is that it minimizes the impact of the Miller capacitor, improving the reverse isolation and stability. In addition, the common-gate topology is widely used in UWB LNA designs for its wideband input match. However, at 60 GHz, the common gate structure cannot provide large gain.
2. Common source (CS) structure
   The CS topology, Fig. 8.6b, is one of the most popular architectures in LNA designs. With the inductive source degeneration to provide the desired real part of the input impedance, the input of the LNA can be matched to 50 Ω. Due to the negative feedback, the NF and linearity performance are also improved slightly. The downside of this architecture is that when the frequency increases, the CS LNA is vulnerable to the Miller effect and hence has a degraded reverse isolation and stability.

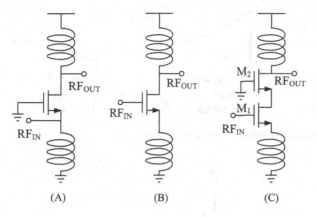

Fig. 8.6 Simplified schematic of different LNA topologies. (**a**) CG; (**b**) CS; (**c**) Cascode

3. Cascode structure

   The cascode structure, Fig. 8.6c, offers conditions for achieving high gain, good isolation, input matching, and low noise simultaneously. The achieved G_a and reverse isolation help to minimize the impact of following stages on the overall NF and S_{11} of the LNA, which is attractive in a multi-stage LNA design. A potential drawback of the cascode architecture is that the noise mismatch between M_1 and M_2 can degrade the total NF of the LNA. Hence, a noise matching network may be necessary between the CG (M_2) and CS (M_1) transistors.

   Based on the discussion above, a cascode structure as first stage and second stage has been chosen. It incorporates the advantages of both CG and CS topologies. To further improve the noise performance, a matching network is inserted between the CG and CS transistors, which will be elaborated in the next section. An inter-stage matching network is used to offer power matching and reasonable noise matching between the first and second stages.

8.4.3 Design Strategy

As shown in Fig. 8.5, the low threshold NMOS transistor has the peak f_{max} with current density at 0.37 mA/μm, while the optimum NF_{min} (60 GHz) is achieved at 0.18 mA/μm. Therefore, a good trade-off is to choose a current density that is around 0.2 mA/μm for the first stage. The proposed LNA is targeting an NF below 4 dB and a transducer gain (G_t) higher than 15 dB. The specifications can be distributed across each stage as shown in Fig. 8.7. Based on Friis noise formula, and assuming identical first and second stage inside the two-stage amplifier, the first stage requires an NF of 3.4 dB and an available gain of 8 dB, and the second state requires an NF of 4 dB and an available gain of 8 dB.

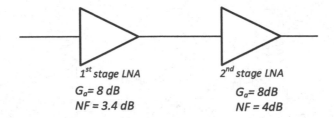

Fig. 8.7 Specification of first and second stage of the LNA

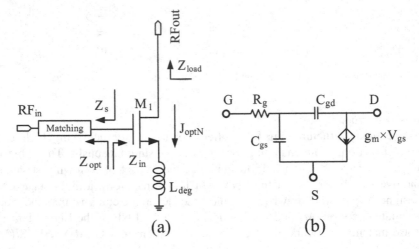

Fig. 8.8 (a) Schematic of a unit-size degenerated CS LNA and (b) the transistor small-signal model

8.4.3.1   Simultaneous Noise and Gain Match

The input matching of the LNA is crucial, and the goal of the input matching is to minimize the degradation of the NF from NF_{min}. At the same time, the goal of conjugate matching is to maximize both G_t and G_a. With the maximized first stage gain G_a, the impact of noise from the later stage can be minimized. However, normally, the optimized noise source impedance Z_{opt} is different from the source impedance Z_{in}. Thus, there is a trade-off between NF and input return loss. The target of the input stage of the LNA design is to make Z_{opt} close to Z_{in}^*, in which case the noise and gain matching can be achieved simultaneously. Mathematically, this is equivalent to minimizing the reflection coefficient [24], Γ_{in-opt}

$$\left|\Gamma_{in-opt}\right| = \left|\frac{Z_{in} - Z_{opt}^*}{Z_{in} + Z_{opt}^*}\right| \tag{8.6}$$

In the unit-size inductively degenerated common-source transistor, Fig. 8.8a, biased at the optimized noise current density, J_{opt}, the optimal noise source impedance, $Z_{\text{opt-unit}}$, and input impedance, $Z_{\text{in-unit}}$, of the transistor can be derived from the Z parameters [25]:

$$Z_{\text{in}} = \frac{1}{N_{\text{finger}}} Z_{\text{in-unit}} = \frac{1}{N_{\text{finger}}} \left(Z_{11} - \frac{Z_{12}Z_{21}}{Z_{22} + Z_{\text{load}}} \right) \qquad (8.7)$$

where $Z_{ij|i,j=1,2}$ can be expressed as

$$z_{ij|i,j=1,2} = z_{ij-\text{unit}|i,j=1,2} + sL_{\text{deg}} \qquad (8.8)$$

The gate series matching inductor can be expressed as

$$L_{\text{gate}} = \frac{Z_0^2 C_{\text{pad}}}{1 + \omega^2 C_{\text{pad}}^2 Z_0^2} - L_{\text{deg}} + \frac{1}{C_{\text{gs}}\omega^2 \left(1 + \omega^2 C_{\text{pad}}^2 Z_0^2 \right)} \qquad (8.9)$$

where Z_0 is the input termination impedance, C_{pad} is the pad capacitance, C_{gs} is the transistor gate-source parasitic capacitance.

Z_{opt} can be expressed as

$$Z_{\text{opt}} = Z_{\text{opt-unit}} - sL_{\text{deg}} \qquad (8.10)$$

where $Z_{ij-\text{unit}|i,j=1,2}$ are the Z-parameters of the unit-size transistor for which the small-signal equivalent circuit is shown in Fig. 8.8b, and Z_{load} is the output termination of the transistor. As a result, with each given transistor size N_{finger} with L_{deg}, an optimum L_{gate}, which leads to simultaneous noise and gain matching, can be found by solving

$$\frac{\partial |\Gamma_{\text{in-opt}}|}{\partial L_{\text{deg}} \partial L_{\text{gate}}} = 0 \qquad (8.11)$$

Based on this derivation, the design should start with the common-source transistor from a minimum size, $W_{\text{finger}} = 1 \, \mu\text{m}$ $L_{\text{finger}} = 40 \, \text{nm}$, and assume that the device is optimally biased at J_{opt}. The Z-parameters for the device can be extracted using the small-signal model in a simulator, and lead to $Z_{\text{opt0}} = 348 + j1945$ at 60 GHz. Based on the current range which is from the IIP$_3$ requirement, the finger number is chosen to be 35. The optimum relation between L_{gate} and L_{deg} can now be derived by substituting (8.9), (8.10) in (8.11) and is plotted in Fig. 8.9. However, with the increased value of the degeneration inductor, the current feedback will decrease the available gain, as shown in the right y-axis of Fig. 8.9. The combination $L_{\text{deg}} = 100 \, \text{pH}$ and $L_{\text{gate}} = 190 \, \text{pH}$ leads to a real part of Z_{in} close to 50 Ω. Figure 8.10 shows the simulated impedance transformation in the Smith chart, providing a more intuitive way to demonstrate how simultaneous noise and gain matching is achieved with L_{deg} and L_{gate}. Starting point are Z_{opt0} and Z_{in0}, marked with the black crosses, which are the optimal noise source impedance and input impedance of the unit-size CS transistor, respectively. By scaling the transistor width 35 times up, both

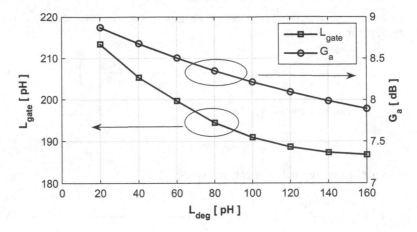

Fig. 8.9 Different L_{deg} with L_{gate} that minimize Γ_{in-opt} and L_{source} with G_a

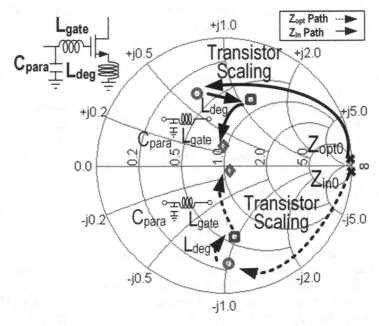

Fig. 8.10 Transformation of Z_{opt} and Z_{in} in the Smith chart by scaling the transistor, degeneration inductor, and matching network

impedances decrease by 35 times. The added 100 pH degeneration inductor L_{deg} increases the real part of Z_{in} and decreases the imaginary part of Z_{opt}. Z_{opt} and Z_{in} become almost conjugate to each other, shown by the squares in Fig. 8.10. Finally, with a series 200 pH inductor at the input, both noise and gain matching can be achieved for 50 Ω source impedance.

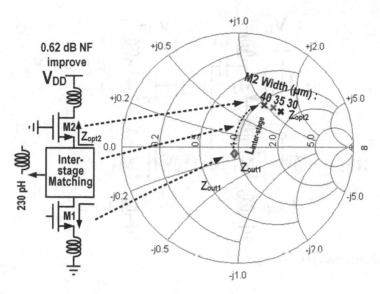

Fig. 8.11 Design of the inter-stage noise matching network between the common source stage and the common gate stage

8.4.3.2   Noise Matching Between Cascode Transistors

In the cascode structure, the noise from the common gate stage will degrade the overall NF. Therefore, a proper noise matching is required between the common source and the common gate stage. The matching network is located between the output of the common source stage and the input of the common gate stage. It is used to transform the output impedance of the common source stage, Z_{out1}, to the noise source impedance of the common gate stage, Z_{opt2} (Fig. 8.11). The width of M_2 is determined at the same time keeping the current density to J_{opt}. By increasing the finger number of M_2, Z_{opt2} moves close to the low impedance side of the Smith chart as shown in Fig. 8.11. To simplify the inter-stage matching, the length of M_2 is chosen as 40 nm, so the real part of Z_{opt2} is close to the real part of Z_{out1}. In this case, the matching can be done by a series inductor. Compared to the direct connection between the common source and the common gate stage, the proposed structure with noise matching between them can improve the NF by 0.08 dB at 60 GHz, which is important to reach the target NF of 3.4 dB for the first stage.

The complete two-stage LNA schematic is shown in Fig. 8.12, showing the input matching for the first stage (L_1), and the inter-stage matching between the common source and common gate stage (L_2, L_5). L_4 achieves the inter-stage matching between the first stage and the second stage. The output matching to 50 Ω is formed by L_6 and L_7.

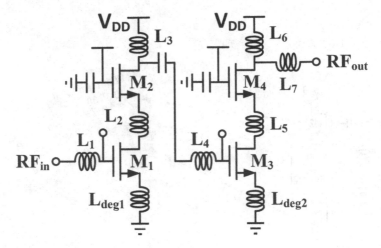

Fig. 8.12 Schematic of two-stage LNA

8.4.4   Measurement Result

The chip photo of the LNA is shown in Fig. 8.13 with a chip area of $0.63 \times 0.31\,\text{mm}^2$. Because there is no thick metal in this CMOS 40 nm technology, the top two metal layers are stapled together with the AP layer. The measured input return loss and transducer gain are shown in Fig. 8.14. The LNA has the peak G_t of 15 dB at 55.1 GHz and 12.5 dB at 60 GHz. The 3 dB power bandwidth covers 48 GHz to 61 GHz. The input return loss is better than 10 dB in a bandwidth around the frequency of interest. The measured NF is shown in Fig. 8.15. The best NF is 3.6 dB at 55 GHz and 3.8 dB at 60 GHz. The total power consumption is 20.4 mW.

8.4.5   Conclusion

In this section, the design of a two-stage 60 GHz LNA is presented using a 40 nm CMOS technology. To achieve optimum performance, different LNA topologies are compared in the 60 GHz band. The proposed LNA employs simultaneous noise and input power matching techniques and noise matching between the CS-CG transistors, thereby minimizing the degradation of the noise performance. A measured NF of 3.8 dB is achieved at 60 GHz. Table 8.2 shows the benchmark with other silicon-based LNAs operating in the 60 GHz ISM band. The proposed work has an improved NF over recently published silicon-based LNAs.

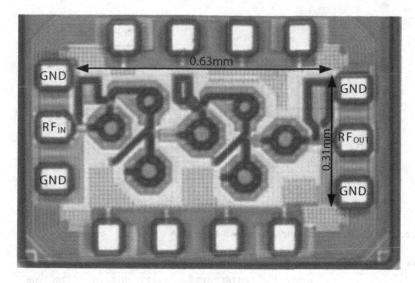

Fig. 8.13 Die photo of proposed two-stage LNA

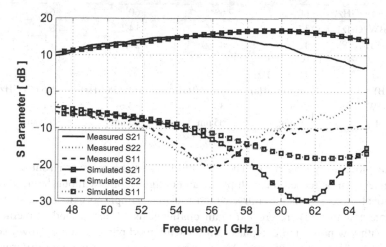

Fig. 8.14 Simulated and measured S-parameter of two-stage LNA

8.5   60 GHz 5-Bit Digitally Controlled Phase Shifter

Phased-array systems can improve the SNR of the RX chain. At 60 GHz, the 5 mm wavelength in silicon makes it possible to integrate antenna arrays on chip for both TX and RX. By using directional high gain antennas with narrow beam width, beam steering is enabled electronically by on-chip phase shifters. The RF path phase shifter architecture offers a solution of small area with relaxed system linearity requirement. In the RF path phase shifter architecture, only one mixer is required

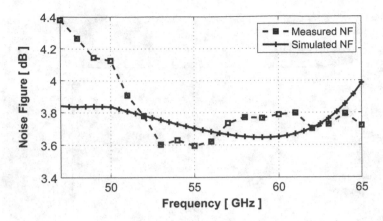

Fig. 8.15 Simulated and measured NF of two-stage LNA

Table 8.2 Performance comparison table of 60 GHz LNAs

	This work	[26] 2012 TMTT	[27] 2011 RFIC	[28] 2008 JSSC
Frequency (GHz)	60	60	60	64
3-dB BW (GHz)	13	12	14.1	8
Gain (dB)	12.5 (60 GHz)	8	20.6	15.5
	15 (56 GHz)			
NF (dB)	3.6 (55 GHz)	4.5 (60 GHz)	4.9 (58 GHz)	6.5 (64 GHz)
P_{diss} (mW)	20.4	35	33.6	86
Technology	40 nm	65 nm	65 nm	90 nm

for the entire array (Fig. 8.2a). The core circuitry of the receiver (front-end, up to the mixer) can be reused for multiple array configurations, without the need to add additional mixers to the circuitry, which makes it simple to extend the system to multiple antennas. Therefore the main challenge is to design and implement an LNA with low noise figure, a phase shifter with good phase accuracy, low loss and small phase/loss deviation and a VGA as compensation for the gain difference of the phase shifter. In this section, a 60 GHz passive 5-bit digitally controlled phase shifter is introduced with low loss and small phase deviation.

8.5.1 Phase Shift Realization

A passive phase shifter can be realized as a high pass or low pass filter, as shown in Fig. 8.16 [22]. The phase shifter could be realized by a high pass or a low pass filter. Each of them could be implemented in Π or T structure. The component values for realizing the high pass filter can be calculated by (8.12)–(8.15) [22], and the

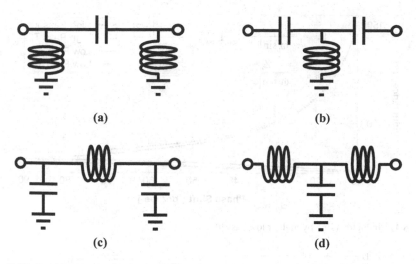

Fig. 8.16 Passive phase shifter realized by high or low pass filters. (**a**) High pass filter: π type. (**b**) High pass filter: T type. (**c**) Low pass filter: π type. (**d**) Low pass filter: T type

values for realizing the low pass filter are given by (8.16)–(8.19) [22], where Z_0 is the characteristic impedance at the ports and ω_0 is the center frequency.

$$L_{high\_\pi} = \frac{Z_0}{\omega_0 \tan |\varphi/2|} \tag{8.12}$$

$$C_{high\_\pi} = \frac{1}{\omega_0 Z_0 \sin |\varphi|} \tag{8.13}$$

$$L_{high\_T} = \frac{Z_0}{\omega_0 \sin |\varphi|} \tag{8.14}$$

$$C_{high\_T} = \frac{1}{\omega_0 Z_0 \tan |\varphi/2|} \tag{8.15}$$

$$L_{low\_\pi} = \frac{Z_0 \sin |\varphi|}{\omega_0} \tag{8.16}$$

$$C_{low\_\pi} = \frac{\tan |\varphi/2|}{\omega_0 Z_0} \tag{8.17}$$

$$L_{low\_T} = \frac{Z_0 \tan |\varphi/2|}{\omega_0} \tag{8.18}$$

$$C_{low\_T} = \frac{\sin |\varphi|}{\omega_0 Z_0} \tag{8.19}$$

The component value, L and C, will depend on the chosen structure. Those differences could cause robustness issues from the process variation. The values

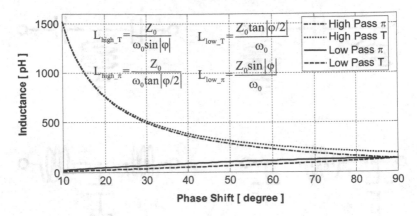

Fig. 8.17 Inductor value by high or low pass filters

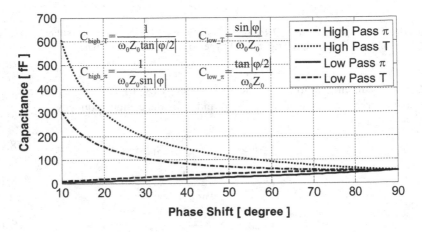

Fig. 8.18 Capacitor value by high or low pass filters

for inductors and capacitors for the low or high pass filter are calculated and summarized in Figs. 8.17 and 8.18 for a phase shift smaller than 90°, where φ is the desired phase shift, Z_0 is the characteristic impedance, and ω_0 is the center frequency.

The different components values are also plotted in Figs. 8.17 and 8.18 as a comparison. In a low pass filter structure, the value of L and C is increasing with frequency, while it is decreasing with frequency in a high pass filter. However, the L value is much larger in the high pass filter for small φ, which makes it difficult to implement in CMOS technology with a reasonable performance. Considering the technology implementation issues and the parasitic capacitors, it is better to apply a low pass filter for small φ instead of a high pass filter. On the other hand, the passive components are too small if using low pass filters. This would make the design more sensitivity to process variation. For example, to provide 11.25° phase

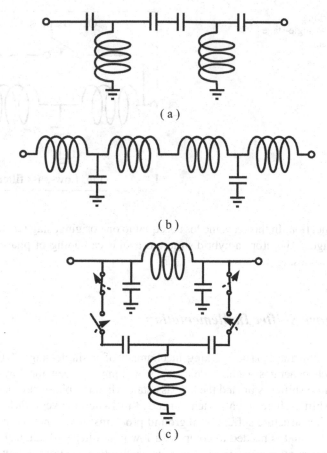

Fig. 8.19 Passive phase shifter realized by high or low pass filters for 180° phase shift. (**a**) Two stage series connected high pass filter. (**b**) Two stage series connected low pass filter. (**c**) Hybrided parallel connected low pass filter and high pass filter

shift, the required value for inductor and capacitor are 25.9 pH and 5.2 fF (low-Π), 13 pH and 10 fF (low-T). Although low-Π network has the advantage of only one inductor (area), the capacitor is too vulnerable to process variation. For low-T networks, as long as the two closely located inductors would not be too big to suffer from inductor coupling, T networks are the better choice for implementing all the four networks.

The hybrid structure, shown in Fig. 8.19, is composed of high pass and low pass filter. The total phase shifting is the difference between two paths. In the 180° situation, the phase achieved is the difference between 90° phase delay from the low pass filter and the 90° phase ahead from the high pass filter. In this way, it can take advantage of the different frequency properties of high pass and low pass filter, and the whole frequency response is much more flat and broadband. Also compared with the cascade of two identical high/low pass structures, a hybrid structure provides a

Fig. 8.20 Working mechanism of a single-stage T-type phase shifter

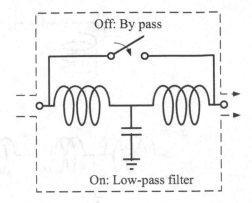

parallel connection. In this way, the loss is equal to one original stage instead of two cascaded stages. Therefore a hybrid structure avoids cascading of phase deviation and passive losses.

8.5.2   Phase Shifter Implementation

Figure 8.20 illustrates basic working mechanism of a single-stage T-type phase shifter, which generates a phase delay by switching between the low-pass state when the phase shifter is on and the by-pass state when the phase shifter is off with zero phase shifts. Therefore a switch is needed to switch between different states. Besides, such a structure suffers from ground problems: when the low-pass state is on, an ideal ground is needed to compose a low pass filter with an accurate delay, while the by-pass state is on, an ideal open is needed to get rid of the low pass filter. Fortunately, an LC tank with a capacitor implemented by a transistor could solve this problem. Additionally, the capacitor that forms the low pass filter could also be implemented by transistors which suffer less from process variation, at least in this specific IC process.

Figure 8.21 shows the schematic of a single-stage T-type switched phase shifter and its equivalent circuits for different states. The circuit is implemented with transistors as switches and a large resistor is added to the gate of each transistor to isolate the RF signal from the control bias signal. When the low-pass state is on, transistor Q_1, Q_2 are turned off and Q_3 is turned on, the equivalent circuit forms a low pass filter and transistor Q_3 provides a path to ground with a small on-resistance R_{on}. When the by-pass state is on, transistors Q_1, Q_2 are turned on and Q_4 is turned off, the equivalent circuit forms a by-pass state. The parasitic capacitance of transistor Q_3 and L_2 forms an LC resonator, providing an open port. Therefore, these two equivalent circuits will provide the desired phase shift in the frequency band of operation.

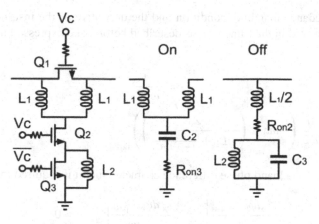

Fig. 8.21 Schematic of a single-stage T-type phase shifter

When the input/output port is matched to $Z_0 = 50\,\Omega$, the inductance and capacitance required to provide the insertion phase delay φ are set by

$$L_1 = \frac{Z_0}{\omega_0}\tan\left(\frac{\varphi}{2}\right) \tag{8.20}$$

$$C_2 = \frac{\sin(\varphi)}{\omega_0 Z_0} \tag{8.21}$$

Thus the derivative of the insertion phase at frequency ω_0 can be expressed by

$$\frac{d}{d\omega}\left(-\tan^{-1}\left(\frac{Z_0\omega C_2}{2-\omega^2 L_1 C_2}\right) + \tan^{-1}\frac{\omega L_1}{Z_0}\right)\Bigg|_{\omega=\omega_0} = \frac{-2L_1}{Z_0} \tag{8.22}$$

When the by-pass state is on, the input admittance of the shunt elements can be written as

$$Y_{\text{in}} = -j\frac{1}{\frac{\omega L_1}{2} + \frac{\omega L_2}{1-\omega^2 L_2 C_3}} \tag{8.23}$$

Therefore the transmission matrix for the shunt elements is given by

$$\begin{vmatrix} A & B \\ C & D \end{vmatrix} = \begin{vmatrix} 1 & 0 \\ Y_{\text{in}} & 1 \end{vmatrix} = \begin{vmatrix} 1 & 0 \\ -j\frac{1}{\frac{\omega L_1}{2} + \frac{\omega L_2}{1-\omega^2 L_2 C_3}} & 1 \end{vmatrix} \tag{8.24}$$

S_{21} and S_{11} can be described as

$$S_{21} = \frac{2}{A + \frac{B}{Z_0} + CZ_0 + D} \tag{8.25}$$

$$S_{11} = \frac{A + \frac{B}{Z_0} - CZ_0 - D}{A + \frac{B}{Z_0} + CZ_0 + D} \tag{8.26}$$

The impedance matching condition and the derivative of the insertion phase at ω_0 can be derived in the same way as described before, and expressed in (8.27) and (8.28)

$$L_2 = \frac{1}{\omega^2 C_3} \tag{8.27}$$

$$\frac{d}{d\omega}\left(-\tan^{-1}\left(\frac{Z_0}{\omega L_1 + \frac{2\omega L_2}{1-\omega^2 L_2 C_3}}\right)\right)\Bigg|_{\omega=\omega_0} = -C_3 Z_0 \tag{8.28}$$

To yield a broadband phase difference characteristic, it is required that

$$\frac{d\varphi_{\text{low-pass}}}{d\omega}\Bigg|_{\omega=\omega_0} - \frac{d\varphi_{\text{by-pass}}}{d\omega}\Bigg|_{\omega=\omega_0} = 0 \tag{8.29}$$

$$\frac{-2L_1}{Z_0} = \frac{d\varphi_{\text{low-pass}}}{d\omega}\Bigg|_{\omega=\omega_0} = \frac{d\varphi_{\text{by-pass}}}{d\omega}\Bigg|_{\omega=\omega_0} = -C_3 Z_0 \tag{8.30}$$

$$C_3 = \frac{2L_1}{Z_0^2} = \frac{2\tan\left(\frac{\varphi}{2}\right)}{\omega_0 Z_0} \tag{8.31}$$

Therefore the equivalent circuit elements are expressed as functions of ω_0 and φ. With the assumption of $Z_0 = 50\,\Omega$ and $\omega_0 = 2\pi \times 60\,\text{GHz}$, the phase difference characteristics versus frequency of $11.25°$ stage are plotted in Fig. 8.22, with a minimum phase variation around $60\,\text{GHz}$. Figure 8.23 compares the insertion phase flatness of $180°$ stages using cascade low-pass structure and the hybrid structure mentioned before. The hybrid structure tends to have smaller phase variation than the cascaded structure.

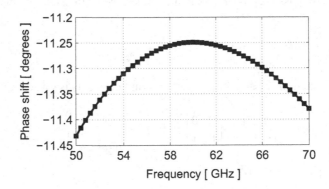

Fig. 8.22 Phase difference of $11.25°$ stage using T-type phase shifter

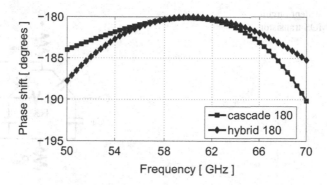

Fig. 8.23 Phase difference comparison of 180° stage

8.5.2.1 CMOS Switches

The switches inside the phase shifter are implemented by NMOS transistors. A large bias resistor is added at the gate of this NMOS transistor to prevent signal leakage and limit oxide breakdown. The insertion loss of the transistor-based switch is affected by the source/drain junction capacitance, the on-state resistance and the coupling to the substrate. In 40 nm CMOS technology, the NMOS transistor can be fabricated in deep n-well structure. In the deep n-well technology, the extra n-well is implanted between the p-type silicon and the transistor. In this way, there is a body contact to bias the bulk. A separate bulk can be connected to the source instead of the silicon p-substrate to reduce the insertion loss. Body-source connected and body-grounded transistor-based switches are analyzed and compared. The equivalent circuit of a CMOS transistor with deep n-well technology is shown in Fig. 8.24. R_{bias} (typically 20 kΩ) is the biasing resistor. The insertion loss arises from the coupling of the drain and the source nodes through junction capacitor C_{db} and C_{sb} to ground at high frequencies and the on-resistor R_{ds} at low frequencies. When the source and body are connected, the insertion loss decreases as the ground path through the substrate is removed. On the other hand, when the switch is turned off, the drain and source are connected directly through C_{db} since the source and body are connected together, which degrades the isolation of the switch. The on-state resistance could be reduced by increasing the size of the transistors, which also introduces larger junction capacitors resulting in more loss and poorer isolation.

Figure 8.25 illustrates the simulation results for the insertion loss for body-grounded and body-source connected switches and it is clear that the insertion loss has a 0.5 dB improvement at 60 GHz when the body-source connected transistor is used. From simulation it is shown that there is an optimum value for the transistor size to minimize the insertion loss while the isolation performance is decreasing as the transistor becomes wider. Therefore, the transistor switch with body-source connection could improve the insertion loss at the cost of isolation.

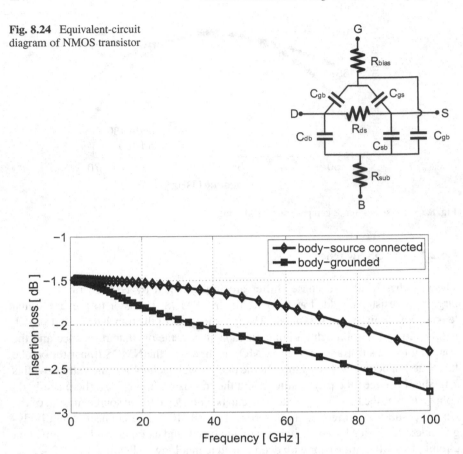

Fig. 8.24 Equivalent-circuit diagram of NMOS transistor

Fig. 8.25 Comparison of NMOS transistor with body-source and body-ground connection

8.5.2.2 Sequence of Phase Shifting Stages

The phase shifter is implemented with five stages with the finest resolution of 11.25°, corresponding to 3.58° in space. The sequence of the phaser shifter is 180°, 22.5°, 45°, 90°, and 11.25°. This sequence of the phase shifter is taken into account of the loading effect and inter-stage matching. As calculated before, small phase shifting stages have smaller inductors and capacitors and thus are more sensitive to mismatches and loading effects than large phase shifting stages. Therefore small phase shifting stages such as 11.25° stage and 22.5° stage are located in between large phase shifting stages to reduce the influence of loading and have better phase linearity, as shown in Fig. 8.26.

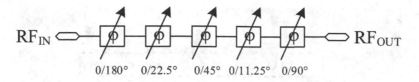

Fig. 8.26 Sequence of stages of the proposed 5-bit switch-type phase shifter

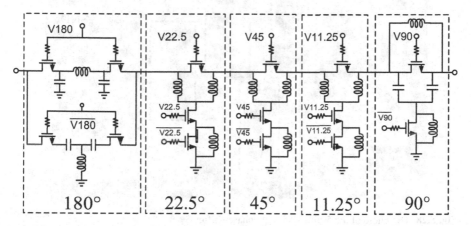

Fig. 8.27 Schematic of the 5-bit switch-type phase shifter

8.5.3   Phase Shift Schematic and Layout

The 5-bit phase shifter is designed using five stages of switch-type delay networks. The 180° stage is realized by switching between a low pass and a high pass network. The following 22.5°, 45°, and 11.25° stages are implemented by using low-pass T-type networks, switching between low-pass state and by-pass state. The 90° stage is implemented by a Π-type network, because the inductors are large and located nearby, they will suffer from coupling problems. For low insertion loss, the transistor switches are implemented with body-source connection. The schematic of the proposed 5-bit 60 GHz phase shifter is shown in Fig. 8.27.

8.5.4   Measurement Results and Comparison to State-of-the-Art

The chip photo of the proposed 5-bit switch type RF phase shifter is shown in Fig. 8.28 with a chip area of $0.6 \times 0.4\,\text{mm}^2$ excluding RF and DC pads. Two-port on-wafer S-parameter measurements were conducted in a frequency range from 50 GHz to 65 GHz using a vector network analyzer (Agilent N5247A). A total of

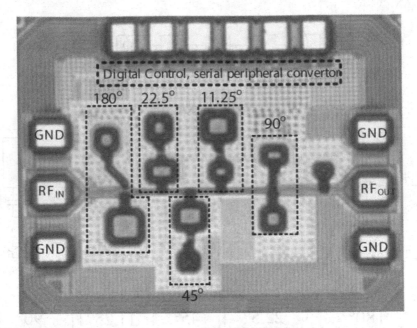

Fig. 8.28 Die photo of the 5-bit switch-type phase shifter

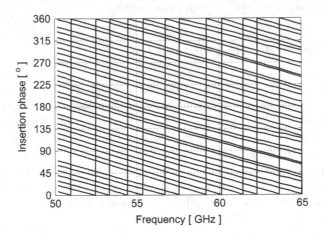

Fig. 8.29 Measured insertion phase of 32 different phase states as a function of frequency

32 combinations of the bias conditions are set through the on-chip series-to-parallel convertor to measure the 32 phase shift states.

The measured insertion phase of the 32 phase settings is depicted in Fig. 8.29. The phase step is approximately 11.25°. Figure 8.30 highlights the relative phase shifts for 32 phase settings by setting the phase state 00000 as a reference. This shows that the 5-bit phase shifts achieved are relatively constant over a wide

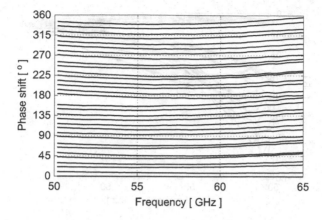

Fig. 8.30 Measured relative phase shift of 32 different phase states as a function of the frequency

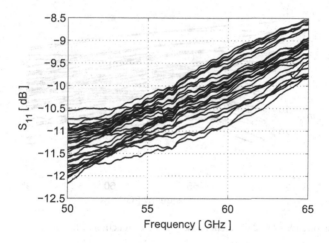

Fig. 8.31 Measured S_{11} of 32 different phase states as a function of frequency

frequency range. Figures 8.31 and 8.32 show the input and output return loss for the 32 phase settings. Figure 8.33 shows the measured insertion loss of the 32 phase shifting states over the frequency range 50–65 GHz . The measured insertion loss achieves loss flatness of ± 1 dB through the 57–64 GHz band. Derived from the measured insertion phase and gain shifts, Fig. 8.34 shows the RMS phase errors and Fig. 8.35 shows the RMS gain errors of the 32 phase states. The RMS phase error is 5° and the RMS gain errors are 2 dB at 60 GHz.

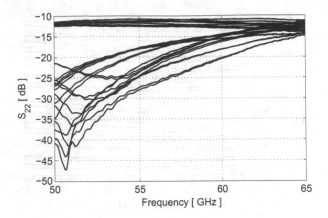

Fig. 8.32 Measured S_{22} of 32 different phase states as a function of frequency

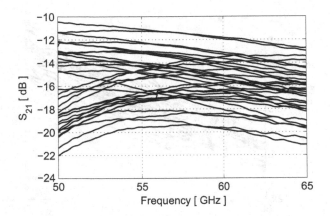

Fig. 8.33 Measured S_{21} of 32 different phase states as a function of frequency

8.5.5 Conclusion

This section presents a state-of-the-art digitally controlled passive phase shifter design in 40 nm CMOS technology. The analysis includes the technology limitations, circuit trade-offs, and the implementation. This digitally controlled phase shifter is a passive phase shifter, which means it can achieve low power consumption when integrated with the LNA in an RF front-end. Table 8.3 summarizes the simulated circuit performance and compares it to recently published RF phase shifters. Compared with the other RF phase shifters, the proposed 5-bit phase shifter has more phase shifting stages, 32 states, and more linear phase shifting performance for the entire frequency range while achieving a comparable loss.

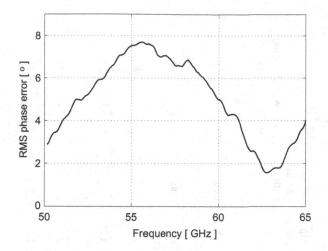

Fig. 8.34 Measured RMS phase errors of the phase shifter

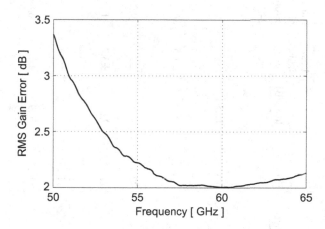

Fig. 8.35 Measured RMS gain errors of the phase shifter

8.6 Conclusion

In this chapter, a phased-array architecture for base stations is proposed. In a wireless power transfer system, the phased-array system could increase the power density at the sensor node by using beam-forming techniques, which could increase the efficiency of energy conversion. In the wireless communication, the phased-array system could increase the SNR for Rx. Compared with IF, LO, and digital baseband phase shifting, in the RF-phase shifting architecture, the signal combining and phase shifting are carried out at RF, which could relax noise figure and linearity requirements of following blocks and allow compensation for other system requirements such as power consumption. The LNA and RF phase shifter are the

Table 8.3 Performance comparison table of RF phase shifters

Ref./Technology	Freq. (GHz)	Topology	Phase range(°)	Resolution (°)	Max. loss(dB)	Loss flatness (dB)	Max. phase error(°)	Min. phase error(°)	Gain deviation (dB)	P_{DC}(mW)	Area(mm²)
[29]/0.13 μm SiGe BiCMOS	32–36	STPS	360	22.5	15	±2.5	24	4	1	0	0.21
[15]/0.18 μm SiGe BiCMOS	40–45	VS	360	n/a	12.5	n/a	9	6.5	n/a	40	0.11
[16]/0.13 μm SiGe BiCMOS	57–64	RTPS	180	n/a	8	±1.5	n/a	n/a	n/a	0	0.56
[1]/65 nm CMOS	55–65	STPS	360	22.5	16	±2	9.2	5.5	n/a	0	0.2
[13]/0.12 μm SiGe BiCMOS	67–78	STPS	360	22.5	22.5	±3.7	11.25	3.75	4.5	0	0.28
[11]/90 nm CMOS	57–64	STPS	360	11.25	18	±0.8	10	2	1.8	0	0.34
This work/ 40 nm CMOS	57–65	STPS	360	11.25	20.9	±1	6.8	1.5	2	0	0.24

most important components in the RF-phase shifting architecture. In this chapter, a design of a two-stage 60 GHz LNA and a design of a 5-bit digital controlled switch type phase shifter are realized using 40 nm CMOS technology. The proposed LNA employs simultaneous noise and input power matching techniques and noise matching between the CS-CG transistors, thereby minimizing the degradation of the noise performance. The measured results show the LNA is 3.8 dB NF and 13 dB G_t at 60 GHz. The proposed phase shifter provides 32 states with 11.25° for each step. It achieves 5° RMS phase errors and 2 dB RMS gain errors at 60 GHz.

References

1. Y. Yu, P. Baltus, A. de Graauw, E. van der Heijden, C. Vaucher, A. van Roermund, A 60 GHz phase shifter integrated with LNA and PA in 65 nm CMOS for phased array systems. IEEE J. Solid State Circuits **45**(9), 1697–1709 (2010)
2. H. Friis, A note on a simple transmission formula. Proc. IRE **34**(5), 254–256 (1946)
3. P. Smulders, Exploiting the 60 GHz band for local wireless multimedia access: prospects and future directions. Commun. Mag. IEEE **40**(1), 140–147 (2002)
4. A. Natarajan, S. Reynolds, M.-D. Tsai, S. Nicolson, J.-H. Zhan, D.G. Kam, D. Liu, Y.-L. Huang, A. Valdes-Garcia, B. Floyd, A fully-integrated 16-element phased-array receiver in SiGe BiCMOS for 60-GHz communications. IEEE J. Solid-State Circuits **46**(5), 1059–1075 (2011)
5. H. Gao, K. Ying, M.K. Matters-Kammerer, P. Harpe, B. Wang, B. Liu, W.A. Serdijn, P.G.M. Baltus, 60 GHz 5-bit digital controlled phase shifter in a digital 40 nm CMOS technology without ultra-thick metals. Electron. Lett. **52**(19), 1611–1613 (2016)
6. S. Reynolds, A. Natarajan, M.-D. Tsai, S. Nicolson, J.-H. Zhan, D. Liu, D. Kam, O. Huang, A. Valdes-Garcia, B. Floyd, A 16-element phased-array receiver IC for 60-GHz communications in SiGe BiCMOS, in *2010 IEEE Radio Frequency Integrated Circuits Symposium (RFIC)* (2010), pp. 461–464
7. H. Veenstra, M. Notten, D. Zhao, J. Long, A 3-channel true-time delay transmitter for 60GHz radar-beamforming applications, in *2011 Proceedings of the ESSCIRC (ESSCIRC)* (2011), pp. 143–146
8. X. Guan, H. Hashemi, A. Hajimiri, A fully integrated 24-GHz eight-element phased-array receiver in silicon. IEEE J. Solid-State Circuits **39**(12), 2311–2320 (2004)
9. H. Hashemi, X. Guan, A. Komijani, A. Hajimiri, A 24-GHz SiGe phased-array receiver-LO phase-shifting approach. IEEE Trans. Microwave Theory Tech. **53**(2), 614–626 (2005)
10. B. Wang, H. Gao, K. Ying, M.K. Matters-Kammerer, P. Baltus, A 60 GHz phased array system evaluation based on a 5-bit phase shifter in CMOS technology, in *2016 Symposium on Communications and Vehicular Technologies (SCVT)* (2016), pp. 1–4
11. W.-T. Li, Y.-C. Chiang, J.-H. Tsai, H.-Y. Yang, J.-H. Cheng, T.-W. Huang, 60-GHz 5-bit phase shifter with integrated VGA phase-error compensation. IEEE Trans. Microwave Theory Tech. **61**(3), 1224–1235 (2013)
12. W.-J. Tseng, C.-S. Lin, Z.-M. Tsai, H. Wang, A miniature switching phase shifter in 0.18 μm CMOS, in *Asia Pacific Microwave Conference, 2009 (APMC)* (2009), pp. 2132–2135
13. S.Y. Kim, G. Rebeiz, A 4-Bit passive phase shifter for automotive radar applications in 0.13 μm CMOS, in *Annual IEEE Compound Semiconductor Integrated Circuit Symposium, 2009 (CISC)* (2009), pp. 1–4
14. C.-W. Wang, H.-S. Wu, C.-K. Tzuang, CMOS passive phase shifter with group-delay deviation of 6.3 ps at K-Band. IEEE Trans. Microwave Theory Tech. **59**(7), 1778–1786 (2011)

15. K.-J. Koh, J. May, G. Rebeiz, A millimeter-wave (40–45 GHz) 16-element phased-array transmitter in 0.18-μm SiGe BiCMOS technology. IEEE J. Solid State Circuits **44**(5), 1498–1509 (2009)
16. M.-D. Tsai, A. Natarajan, 60GHz passive and active RF-path phase shifters in silicon," in *IEEE Radio Frequency Integrated Circuits Symposium, 2009 (RFIC)* (2009), pp. 223–226
17. H. Krishnaswamy, A. Valdes-Garcia, J.-W. Lai, A silicon-based, all-passive, 60 GHz, 4-element, phased-array beamformer featuring a differential, reflection-type phase shifter, in *2010 IEEE International Symposium on Phased Array Systems and Technology (ARRAY)* (2010), pp. 225–232
18. K. Ying, H. Gao, D. Milosevic, P. Baltus, A nonlinear transfer function based receiver for wideband interference suppression. J. Sens. **2017**, 15 (2017)
19. B.-W. Min, G. Rebeiz, Ka-Band BiCMOS 4-Bit phase shifter with integrated LNA for phased array T/R Modules, in *IEEE/MTT-S International Microwave Symposium* (2007), pp. 479–482
20. Y.-C. Chiang, W.-T. Li, J.-H. Tsai, T.-W. Huang, A 60GHz digitally controlled 4-bit phase shifter with 6-ps group delay deviation, in *2012 IEEE MTT-S International Microwave Symposium Digest (MTT)* (2012), pp. 1–3
21. D.-W. Kang, H.D. Lee, C.-H. Kim, S. Hong, Ku-band MMIC phase shifter using a parallel resonator with 0.18 μm cmos technology. IEEE Trans. Microwave Theory Tech. **54**(1), 294–301 (2006)
22. B.-W. Min, G. Rebeiz, Single-ended and differential Ka-Band BiCMOS phased array front-ends. IEEE J. Solid State Circuits **43**(10), 2239–2250 (2008)
23. H. Gao, K. Ying, M.K. Matters-Kammerer, P. Harpe, Q. Ma, A. van Roermund, P. Baltus, A 48-61 GHz LNA in 40-nm CMOS with 3.6 dB minimum NF employing a metal slotting method, in *2016 IEEE Radio Frequency Integrated Circuits Symposium (RFIC)* (2016), pp. 154–157
24. M. Byung-Wook, G. Rebeiz, Ka-Band SiGe HBT low noise amplifier design for simultaneous noise and input power matching. *IEEE Microwave Wireless Compon. Lett.* **17**(12), 891–893 (2007)
25. G.D. Vendelin, A.M. Pavio, U.L. Rhode, *Microwave Circuit Design Using Linear and Nonlinear Techniques*. Wiley-Interscience; 2 edition (July 5, 2005)
26. P. Sakian, E. Janssen, A. van Roermund, R. Mahmoudi, Analysis and design of a 60 GHz wideband voltage-voltage transformer feedback LNA. IEEE Trans. Microwave Theory Tech. **60**(3), 702–713 (2012)
27. H.-H. Hsieh, P.-Y. Wu, C.-P. Jou, F.-L. Hsueh, G.-W. Huang, 60GHz high-gain low-noise amplifiers with a common-gate inductive feedback in 65nm CMOS, in *2011 IEEE Radio Frequency Integrated Circuits Symposium (RFIC)* (2011), pp. 1–4
28. S. Pellerano, Y. Palaskas, K. Soumyanath, A 64 GHz LNA with 15.5 dB gain and 6.5 dB NF in 90 nm CMOS. IEEE J. Solid State Circuits **43**(7), 1542–1552 (2008)
29. J. Roderick, H. Krishnaswamy, K. Newton, H. Hashemi, Silicon-based ultra-wideband beam-forming. IEEE J. Solid State Circuits **41**(8), 1726–1739 (2006)

Chapter 9
Conclusions

Abstract Due to its ease of deployment, wireless sensors have been used in a wide range of applications, including security, green building technology, automotive and health care. Most of the state-of-the-art wireless sensors use batteries as a power source. It can be easily calculated that for a building with 1000 wireless sensors installed, which can be very common for a smart building, assuming a battery life of 3 years, on average, batteries need to be replaced every day. Therefore, from both cost and convenience points of view, there is a strong demand for battery-less wireless sensors. To overcome the limitations of state-of-the-art battery-less wireless sensors in size, cost, robustness, and range, we proposed a 60 GHz wireless sensor system with monolithic sensors in this book. In the PREMISS system, the wireless sensors consist of wireless power receiving, sensing and communication functions in a single chip. The sensors have no external components and hence avoid costly IC-interfaces that are sensitive to mechanical and thermal stress.

Here are the conclusions of the book:

- Rectifiers can be used as a wireless power receiver in CMOS technology. The methods of inductor peaking and local threshold voltage modulation can improve the rectifier efficiency. The principle of those methods is to increase the overdrive voltage in order to decrease the energy loss from the transistor. In Chap. 5, three versions of the mm-wave rectifier were implemented in a 65 nm CMOS technology. The first version of the rectifier is a single-stage inductor-peaking rectifier, it achieved 7% efficiency with 4 dBm input power. With extension from a single-stage to a multi-stage rectifier, the second version of a three-stage rectifier can provide 1 V output voltage with 5 dBm input power at 71 GHz. With the method of local threshold voltage modulation, the third rectifier can achieve 1 V output voltage at -7 dBm input power at 52 GHz.
- It is feasible to implement a mm-wave frequency wireless powered sensor node in CMOS technology. The on-chip antenna integration with the sensor nodes shows it is a direction to achieve a low cost, small area and highly integrated sensor node. In Chap. 6, two versions of such sensor nodes were presented. The area of the first sensor node is 1.09 mm^2 and the second sensor node is 1.25 mm^2,

© Springer International Publishing AG 2018

H. Gao et al., *Batteryless mm-Wave Wireless Sensors*, Analog Circuits and Signal Processing, https://doi.org/10.1007/978-3-319-72980-0_9

including the area of on-chip antenna. The weight is around 2 mg. Those sensors were implemented in 65 nm CMOS technology with on-chip wireless power receiver and temperature monitoring function in order to demonstrate the concept of the monolithic sensor node. The measured temperature information is sent back wirelessly.

- The low average power dissipation of the entire Rx front-end for event-driven communication applications can be achieved using the asynchronous duty-cycled wake-up power management method proposed in this work. It can be achieved through the asynchronous IJLO-based duty-cycled WuRx. It has been shown that this approach can be integrated into a low-power and high-speed wireless system and it helps to minimize the average power consumption of such a system efficiently. In Chap. 7, a self-mixer based receiver in a 65 nm CMOS technology was presented. In this architecture, the injection-locked oscillator is applied instead of a PLL so as to save the total power consumption. In this receiver, the total power consumption is 41 mW, which is quite low for low-power applications.

- It is feasible to implement a low-cost high data-rate 60 GHz wireless system using phased-array architecture in the base-station integrated in IC CMOS technology. The RF-phase shifting architecture shows low cost and low power consumption. Chapter 8 presents a low noise amplifier and a passive phase shifter implemented in 40 nm CMOS technology. The proposed LNA employs simultaneous noise and input power matching techniques and noise matching between the CS-CG transistors, thereby minimizing the degradation of the noise performance. The measured results show that LNA has 3.8 dB NF and 13 dB G_t at 60 GHz. The proposed phase shifter provides 32 states with 11.25° phase shift steps. It achieves 5° RMS phase error and 2 dB RMS gain error at 60 GHz.

Index

© Springer International Publishing AG 2018
H. Gao et al., *Batteryless mm-Wave Wireless Sensors*, Analog Circuits
and Signal Processing, https://doi.org/10.1007/978-3-319-72980-0

Printed in the United States
By Bookmasters